D.B.

CALIXARENES IN ACTION

CALIXARENES IN ACTION

Editors

Luigi Mandolini
Università "La Sapienza" and CNR, Roma, Italy

Rocco Ungaro
Università degli Studi di Parma, Italy

Imperial College Press

Published by

Imperial College Press
57 Shelton Street
Covent Garden
London WC2H 9HE

Distributed by

World Scientific Publishing Co. Pte. Ltd.
P O Box 128, Farrer Road, Singapore 912805
USA office: Suite 1B, 1060 Main Street, River Edge, NJ 07661
UK office: 57 Shelton Street, Covent Garden, London WC2H 9HE

British Library Cataloguing-in-Publication Data
A catalogue record for this book is available from the British Library.

CALIXARENES IN ACTION

Copyright © 2000 by Imperial College Press

All rights reserved. This book, or parts thereof, may not be reproduced in any form or by any means, electronic or mechanical, including photocopying, recording or any information storage and retrieval system now known or to be invented, without written permission from the Publisher.

For photocopying of material in this volume, please pay a copying fee through the Copyright Clearance Center, Inc., 222 Rosewood Drive, Danvers, MA 01923, USA. In this case permission to photocopy is not required from the publisher.

ISBN 1-86094-194-X

Printed in Singapore.

CONTENTS

Contributors		vii
Preface		ix
1.	**Introduction** Rocco Ungaro	1
2.	**Molecular Modeling of Calixarenes and their Host-Guest Complexes** Frank C.J.M. van Veggel	11
3.	**Recognition of Neutral Molecules by Calixarenes in Solution and in Gas Phase** Andrea Pochini and Arturo Arduini	37
4.	**Calixarenes in Spherical Metal Ion Recognition** Alessandro Casnati and Rocco Ungaro	62
5.	**Calixarenes as Hosts for Quats** Antonella Dalla Cort and Luigi Mandolini	85
6.	**Calixarene Based Anion Receptors** Paul D. Beer and James B. Cooper	111
7.	**Structural Properties and Theoretical Investigation of Solid State Calixarenes and their Inclusion Complexes** Franco Ugozzoli	144
8.	**Calixarenes in Thin-Film Supramolecular Materials** Andrew J. Lucke and Charles J.M. Stirling	172
9.	**Calixarenes in Self-Assembly Phenomena** Volker Böhmer and Alexander Shivanyuk	203
10.	**Calixarene Based Catalytic Systems** Roberta Cacciapaglia and Luigi Mandolini	241
Subject Index		265

CONTRIBUTORS

Arturo Arduini, Dipartimento di Chimica Organica e Industriale, Università di Parma, 43100 Parma, Italy

Paul D. Beer, Inorganic Chemistry Laboratory, University of Oxford, Oxford OX1 3QR, United Kingdom

Volker Böhmer, Johannes-Gutemberg-Universität, Fachbereich Chemie und Pharmazie, 55099 Mainz, Germany

Roberta Cacciapaglia, Centro CNR sui Meccanismi di Reazione, c/o Dipartimento di Chimica, Università La Sapienza, 00185 Roma, Italy

Alessandro Casnati, Dipartimento di Chimica Organica e Industriale, Università di Parma, 43100 Parma, Italy

James B. Cooper, Inorganic Chemistry Laboratory, University of Oxford, Oxford OX1 3QR, United Kingdom

Antonella Dalla Cort, Centro CNR sui Meccanismi di Reazione and Dipartimento di Chimica, Università La Sapienza, 00185 Roma, Italy

Andrew J. Lucke, Department of Chemistry, University of Sheffield, Sheffield S3 7HF, United Kingdom

Luigi Mandolini, Centro CNR sui Meccanismi di Reazione and Dipartimento di Chimica, Università La Sapienza, 00185 Roma, Italy

Andrea Pochini, Dipartimento di Chimica Organica e Industriale, Università di Parma, 43100 Parma, Italy

Alexander Shivanyuk, Johannes-Guterberg-Universität, Fachbereich Chemie und Pharmazie, 55099 Mainz, Germany

Charles J.M. Stirling, Department of Chemistry, University of Sheffield, Sheffield S3 7HF, United Kingdom

Franco Ugozzoli, Dipartimento di Chimica Generale e Inorganica, Università di Parma, 43100 Parma, Italy

Frank C.J.M. van Veggel, Laboratories of Supramolecular Chemistry and Technology and Mesa$^+$ Research Institute, University of Twente, 7500 AE Enschede, The Netherlands

PREFACE

The interest in calixarenes in the last five years of this decade has continued with the same geometrical progression experienced in the early 1990s, well documented by the authors of previously published books and review articles on this class of compounds. In this sense calixarenes are rather unique in the panorama of synthetic macrocycles, and it is not surprising that they are often considered as important as cyclodextrins, the most popular class of naturally occurring macrocycles.

Among scientists active in the field of calixarenes, and also outside this community, there is a general agreement that the somewhat surprising continuous interest in calixarenes is mainly due to the fact that they were rediscovered at the right time, *viz.* when *supramolecular chemistry* was in its growing phase. There was need and quest of easily available and synthetically versatile building blocks for the construction of receptor molecules of increasing complexity, able to perform specific supramolecular functions. Calixarenes have both these features since they can be obtained in good yields by one-pot synthesis and can be easily and selectively functionalised both on the phenolic OH groups and on the aromatic nuclei. Calixarenes and some simple derivatives have been commercially available since many years. Their basket shaped structure containing a lipophilic cavity made up of aromatic nuclei has attracted the attention of several theoretical and experimental investigators interested in understanding and mastering weak intermolecular forces such as cation/π or CH/π interaction using simple macrocyclic models. The formation of *molecular capsules* through covalent or noncovalent syntheses, is one of the latest developments in this contest. The calixarene cavity has been exploited as an additional binding site for apolar groups in receptors which use the strategy of *multipoint interactions* in the recognition of polyfunctional guests. On the other hand, the conformational properties of calixarenes have been largely exploited to create new shapes and architectures for molecular receptors. The many reactive positions at the upper and lower rims of the calix have been used to attach binding groups in a precise stereochemical arrangement suitable for *cooperative binding* of guest species.

It seemed therefore appropriate to us and the publisher to devote a book to the supramolecular functions of calixarenes rather than emphasising their chemistry and molecular properties, which are authoritatively covered in other books and recent review articles. The title *Calixarenes in Action* was clearly inspired by the last chapter of Gutsche's recent book (*Calixarenes Revisited* in "Monograph in

Supramolecular Chemistry", ed. Stoddart J. F. ,The Royal Society of Chemistry, Cambridge, 1998). It was our intention to expand and update the content of this chapter covering several aspects of the use of calixarenes in *supramolecular chemistry*. Most of the authors of the ten chapters of this book have been involved in calixarenes since a long time and all of them have given important contributions to the supramolecular aspects of calixarene chemistry.

The following topics have been covered: molecular modeling of calixarenes and their inclusion complexes, use of these macrocycles in the encapsulation, detection and separation of ions and neutral molecules, supramolecular materials based on calixarenes, new complex architectures created by noncovalent synthesis and supramolecular catalysis. Particular attention has been paid to the recent literature and even to unpublished results, although reference to older but significant work has also been given. Due to the very rapid expansion of the subject it was not possible to cover all fields in which calixarenes are *in action*. For example, some practical applications of calixarenes *e.g.* as antioxidants, fuel and polymer additives, stabilisers, optical recording or photoresist materials, etc, have not been covered since they are relatively few and not always linked to the supramolecular properties of calixarenes. We have also made arbitrary choices in selecting and organising the topics. For example, just to quote one topic of wide scope, we decided not to have a specific chapter devoted to metal complexes exploiting the co-ordination properties of the calixarene oxygen atoms and prefer to dilute this important topic in several chapters. We apologise if these choices have caused omissions of important work. Notwithstanding these limitations, we think the ten chapters of *Calixarenes in Action* give a realistic and up-to-date picture of the state of the art of the use of calixarenes in *supramolecular chemistry*.

We warmly thank all of the authors of the various chapters for their valuable contributions. We also thank Dr. Roberta Cacciapaglia and Prof. Alessandro Casnati for their help in the editorial work.

July, 1999

Luigi Mandolini
Rocco Ungaro

CHAPTER 1

INTRODUCTION

ROCCO UNGARO

Dipartimento di Chimica Organica e Industriale dell'Università degli Studi, Parco Area delle Scienze 17/A, I–43100, Parma, Italy.

1.1. Historical Notes, Synthesis and Nomenclature

Although this book is mainly devoted to the supramolecular properties of calixarenes, few words of introduction are needed in order to summarise some basic features of these host molecules, which could facilitate the reading of the following chapters to those not familiar with the subject. For a more extensive coverage of the chemistry of calixarenes the reader should consult the two Gutsche's books[1] and recent review articles.[2]

Calixarenes (**I**) are [1_n] metacyclophanes which derive from the condensation of phenols and formaldehyde in different conditions. As such they have been known since long time[3-5] but it is only in the late seventies, when their structures were firmly established both in solution[6] and in the solid state[7] that they became popular in supramolecular chemistry. By looking at CPK molecular models of the cyclic tetramers derived from the condensation of p-alkyl phenols and formaldehyde, Gutsche[6] coined the name "calix[4]arenes", which derives from Latin "*calix*" (Greek χυλιξ) meaning vase, pointing out the presence of a cup-like structure in these macrocycles when they assume the conformation in which all four aryl groups are oriented in the same direction. This conformation is called *cone* and is usually observed in the solid state.[7] The name was extended to larger macrocyclic compounds. A bracketed number between calix and arene specifies the size of the macrocycle and the name of the p-substituent is added to indicate from which phenol the calixarene is derived. The cyclic tetramer obtained from p-*tert*-butylphenol, for example, is named p-*tert*-butylcalix[4]arene. Since the number of compounds has proliferated during the years, it seemed more appropriate to apply the term "calixarene" to the basic macrocyclic structures devoid of substituents.[1] According to this nomenclature the p-*tert*-butylcalix[4]arene (**1**) is named 5,11,17,23-tetra-*tert*-butylcalix[4]arene-25,26,27,28-tetrol. Authors prefer to use the more systematic name in the experimental part of their papers and the short one in the text.

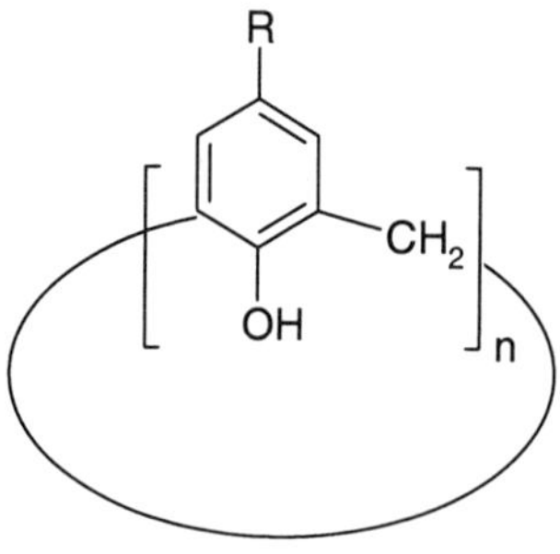

(I) 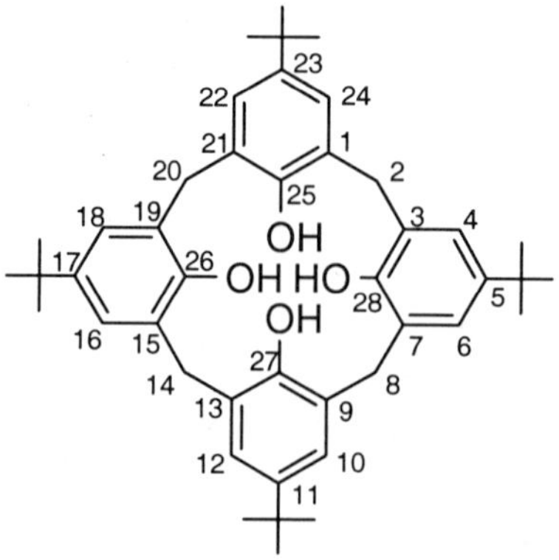   (1)

Two major classes of calixarenes are known, the phenol-derived cyclooligomers (e.g. **1**) and the resorcinol-derived cyclooligomes (e.g. **2**)[8] which are also named (generating sometime confusion) resorcinarenes, resorcarenes, resorcin[4]arenes, calixresorc[4]arenes, depending on the personal preference of the authors.

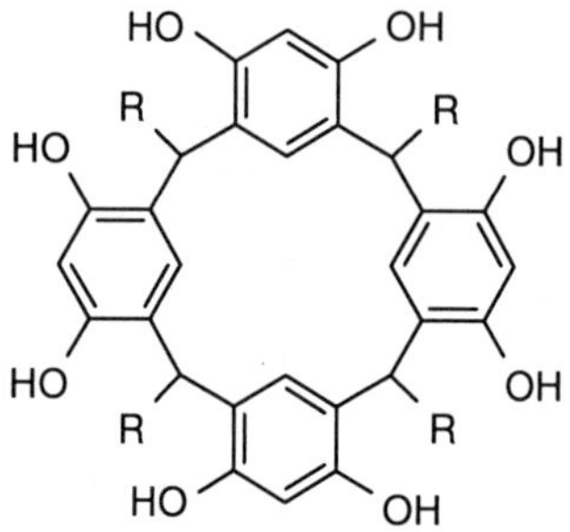

(2) 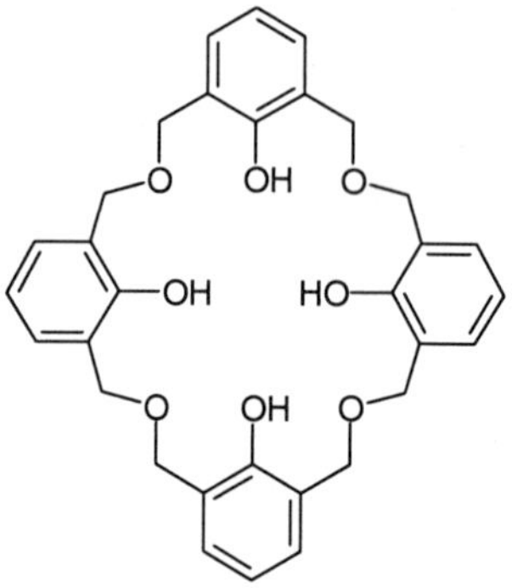   (3)

Apart from the nomenclature there are two main differences between the two classes of calixarenes, the first one is related to the synthesis and the second one to the orientation of the OH groups with respect to the macrocyclic ring. In fact, the calix[n]arenes are usually obtained *via* the base-catalysed condensation of p-alkyl-phenols and formaldehyde and their OH groups are directed towards the interior of the macrocyclic ring (*endo*-OH calixarenes) whereas the calixresorc[4]arenes are prepared through the acid catalysed condensation of resorcinol and aldehydes and are *exo*-OH calixarenes, since these groups are oriented away from the annulus. Most contributions to this book refer to *endo*-OH calixarenes. Also used as receptors or as building blocks are calixarenene-related compounds, the most important being the homooxacalixarenes (e.g. **3**), which can be obtained both by one step and by fragment condensation synthesis.[9]

Two regions can be distinguished in calixarenes (*e.g.* Fig. 1.1), *viz.* the phenolic OH groups region and the para position of the aromatic rings, which are called respectively the "lower rim" and the "upper rim" of the calix. In calix[4]arenes, adjacent nuclei have been named "proximal" or (1,2) whereas the opposite ones are in "distal" or "diametrical" (1,3) positions.

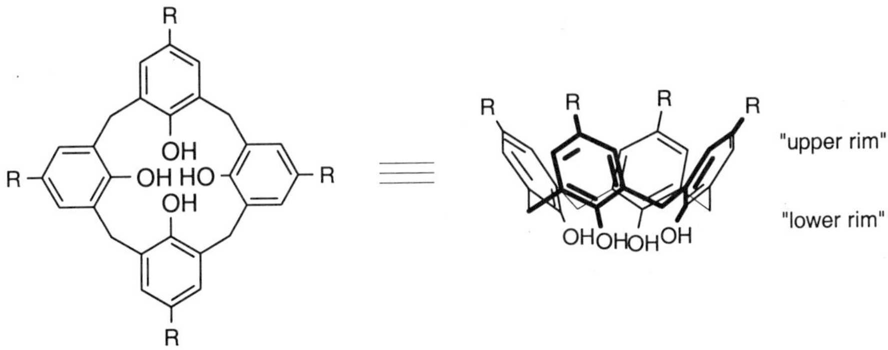

Figure 1.1. The two rims of calix[4]arenes

Although several methods for the synthesis of calixarenes have been developed during the years still the most general and useful is the one-step, base-induced condensations of p-substituted phenols and formaldehyde. With some variations, this method gives good yield of even numbered (n = 4, 6, 8) cyclic products, especially with phenols bearing bulky substituent like *ter*-butyl,[10-12] benzyloxy,[13] and adamanthyl,[14] at *para* position. The odd numbered calixarenes (n = 5, 7, 9) can also be obtained by direct condensation, but yields are considerably lower. A careful investigation, involving changes in base, reactant ratio and reaction temperature, of the one-step condensation between p-tert-butylphenol and formaldehyde by Gutsche's group, resulted in reproducible procedures for the synthesis of p-*tert*-butylcalix[4],[10] [6][11] and [8]arene[12] in good yields. This synthetic procedure has also the advantage that the p-*tert*-butyl group can be easily removed using Lewis or Brönsted acid catalysts to give unsubstituted calix[n]arenes (Scheme 1.1), thus allowing the introduction of other functional groups on the aromatic nuclei.

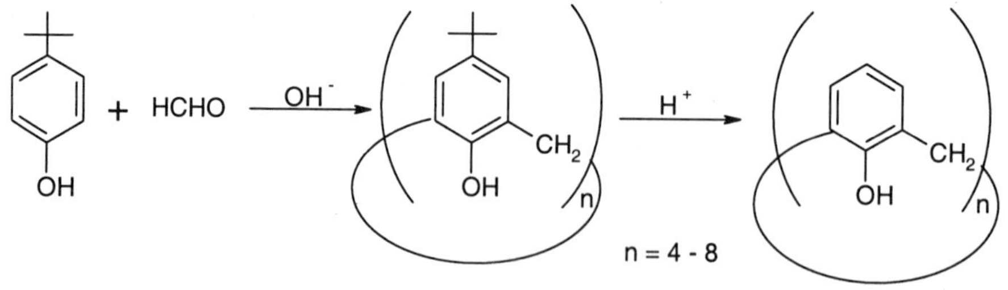

Scheme 1.1. Direct condensation synthesis of p-*tert*-butylcalix[n]arenes and removal of *tert*-butyl groups at the upper rim.

p-*tert*-Butylcalix[6]arene is mainly formed in the presence of large amount of base (KOH or RbOH) whereas high temperatures favour the formation of p-*tert*-butylcalix[4]arene. The cyclic octamer and hexamer can be converted, at high temperature under basic conditions, to the cyclic tetramer.

The one-pot synthesis produces calixarenes having the same substituent at the p-position. Calixarenes with different substituents can be obtained by the stepwise synthesis described by Hayes and Hunter[15] and further optimized by Kämmerer, Happel et al.[16] or, more conveniently, by the convergent stepwise synthesis (fragment condensation) developed by Böhmer and coworkers.[17-18]

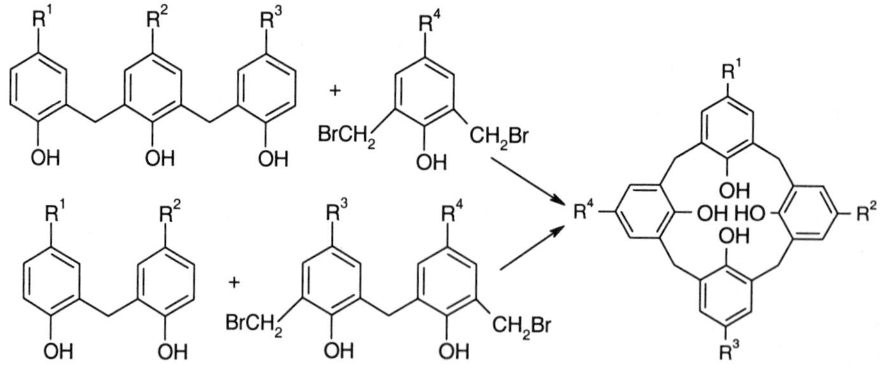

Scheme 1.2. Synthesis of calix[4]arenes by fragment condensation.

In this method (Scheme 1.2) a linear trimer can be condensed with a bisbromomethylated phenol derivative ("3+1" approach[17]) or a suitable linear dimer with a bisbromomethylated dimer ("2+2" approach[18]) to give the desired calix[4]arene derivative in 10-25% overall yields. Special calix[4]arenes have been synthesized by this method.[1-2]

1.2. Conformational Properties and Nomenclature

The ^{1}H NMR spectra of unmodified calix[4]arenes show a pair of doublets at low temperature and a singlet at high temperature for the bridging methylenes. This behaviour has been interpreted as due to the interconversion between two mirror-image *cone* conformations (Fig. 1.2), which is slow (on the NMR time scale) at low temperature and fast at higher temperature.[16] A Nuclear Overhauser Effect investigation has allowed to establish, that the low temperature spectra correspond to the *cone* conformation and that the low field doublet is due to the axial proton H_a whereas the high field doublet is attributed to the equatorial proton H_e.[19] The rate of conformational inversion depends on the substituents at the para-position of the calix only slightly and much more on the solvent. The lowest value of $\Delta G^{\neq}$ at the coalescence temperature (Tc) has been found for calix[4]arenes in pyridine (13.7 kcal mol^{-1}), thus indicating that weakening of intramolecular hydrogen bonding by the polar solvent enhances the rate of ring inversion. In CDCl$_3$ the $\Delta G^{\neq}$ for the same process is 15-16 kcal mol^{-1} for simple p-alkyl calix[4]arenes.[20]

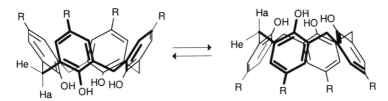

Fig. 1.2. Ring inversion of calix[4]arenes.

Tetramethoxy and tetraethoxycalix[4]arenes are also conformationally mobile but the introduction at the lower rim of calix[4]arenes of alkyl group bulkier than ethyl, blocks the ring inversion process thus producing compounds with different stereochemistry, depending on the orientation of each aryl group which can project upward ("u") or downward (d) relative to an average plane defined by the methylene bridges. The four different possible conformations (Fig. 1.3) have been named by

Gutsche as *cone* (u,u,u,u), *partial cone* (u,u,u,d), *1,3-alternate* (u,d,u,d), *and 1,2-alternate* (u,u,d,d).[1]

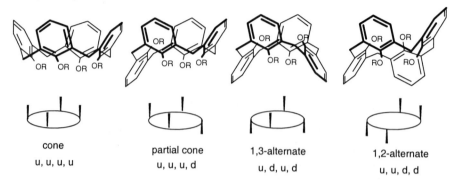

Figure 1.3. Conformations of calix[4]arene derivatives.

They can be easily identified through their ^{1}H and ^{13}C NMR spectra. Particularly useful are the ^{1}H and ^{13}C NMR patterns of the bridging methylene groups which are different for three out of four conformations (Fig. 1.4). The less common *1,2-alternate* conformation shows a similar pattern to the *partial cone*, although the two conformations can be distinguished in the aromatic part of the spectrum.

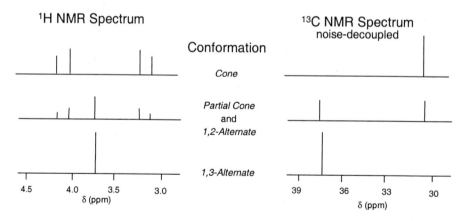

Fig. 1.4. Patterns of the signals expected in the ^{1}H and ^{13}C NMR spectra of the four conformational isomers of calix[4]arenes.

De Mendoza and coworkers have introduced a very useful "rule" for correlating the ^{13}C NMR spectra of calixarenes with their conformation.[21] They have shown that the resonance arising from the bridge methylene carbon is near δ 31 when two adjacent aryl groups are in the *syn* orientation (*i.e.* both "up" or both "down") and near 37 when they are in the *anti* orientation (*i.e.* one group "up" and the other group "down"). The de Mendoza rule has been applied with success not only to calix[4] but also to calix[5]- and calix[6]arenes.

With the exception of calix[8]arenes, which show a peculiar conformational behaviour,[22] the increase in the number of phenolic units in the cyclic array, increases the conformational mobility of calixarenes in solution, and allows the macrocycles to adopt new shapes. Calix[5]arenes can be conformationally blocked by functionalization at the lower rim,[23] but this becomes more difficult for calix[6]-[24] and calix[8]arenes,[25] although few examples have been reported for both cases. As the number of aryl groups in the macrocycle increases, also the number of possible stereoisomers increases and their designation and representation become difficult. By examining the numerous X-ray crystal structures of calix[6]arene derivatives it appeared soon evident that there were several "departures" from the true "up/down" orientations and from the *cone, partial cone,* etc. conformations. Therefore several other designations and iconographic representations have been introduced in order to convey the structural information in the simplest and (hopefully) most accurate way. For example the solid state conformation of p-*tert*-butyl-calix[6]arene[26] is often referred to as *pinched cone* or *winged* and that of p-*tert*-butyl-calix[8]arene[27] as *pleated loop* (Fig. 1.5).

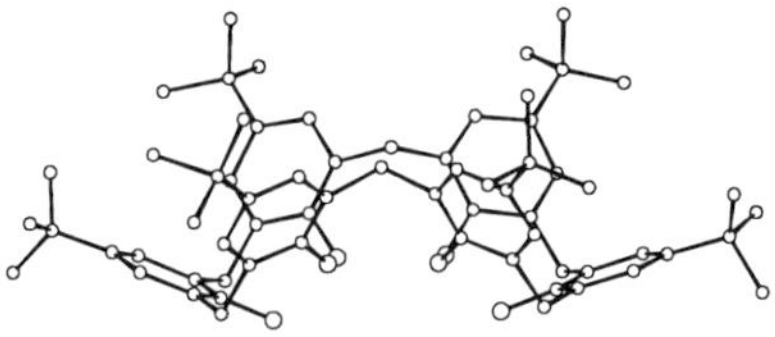

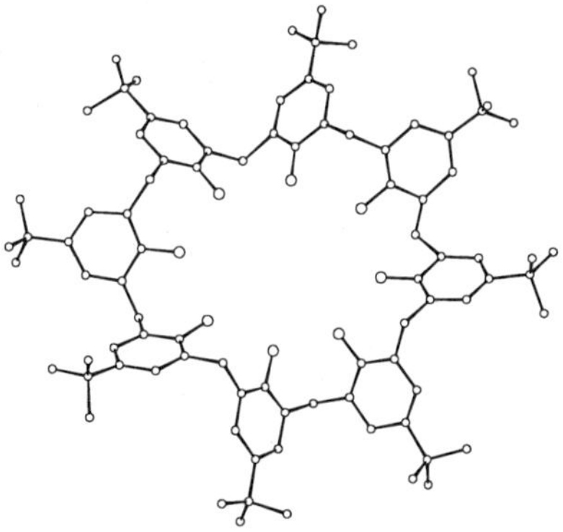

Figure 1.5.   The *pinched cone (winged)* conformation of *p-tert*-butyl-calix[6]arene (left) and the *pleated loop* conformation of p-*tert*-butylcalix[8]arene (right), in the solid state.

Taking into account that the calix[6]arene aryl rings are often not truly perpendicular but have an inclination of 45° or more, Gutsche added the letters "o" (outward) and "i" (inward) to take into account these deviations, thus extending the "up/down" nomenclature to more complex systems.[28] Few representative examples taken from reference 28 and referring to equally substituted calix[6]arene derivatives, are shown in Fig. 1.6.

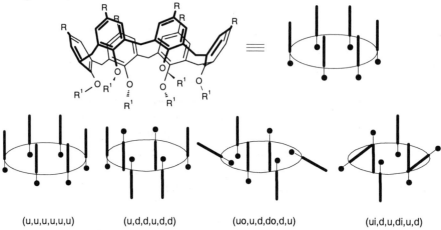

(u,u,u,u,u,u) (u,d,d,u,d,d) (uo,u,d,do,d,u) (ui,d,u,di,u,d)

Figure 1.6. Pseudo-three-dimensional and linear representation of calix[6]arene conformations according to Gutsche.[28]

A more general representation of calixarene conformation was proposed by Ugozzoli and Andreetti,[29] who used as its starting point two adjacent phenolic units A and B linked by a CH$_2$ group, with B lying in the same plane as A and on his right hand side.

They observed that any mutual orientation of the two phenolic rings can be unequivocally obtained by adjusting by ϕ and χ degrees respectively the two torsion angles C(1)-C(2)-C(3)-C(4) and C(2)-C(3)-C(4)-C(5). When the two adjacent rings are oriented *syn*, the signs of ϕ and χ are opposite ; when they are in *anti* orientation the sign of ϕ and χ are the same.

For the precise designation of a calixarene conformation the values as well the signs of all these dihedral angles must be given, but for a qualitative linear designation only the signs need to be given. In cases where high symmetry is observed the Schönflies point symmetry designation can help to simplify the notation. According to this symbolic representation a calix[4]arene *cone* conformation is indicated, in all cases, as + -,+ -,+ -,+ - and C_4 (+ -) if the *cone* is regular (fourfold symmetry) and C_1 (+ -)$_4$ if the *cone* is completely distorted. The *pleated loop* conformation of p-*tert*-butyl-calix[8]arene (Fig. 1.5, right) should be indicated as - +,+ -, - +,+ -, - +, + -, - +,+ - or C_1 (- +, + -)$_4$ and the p-*tert*-butyl-calix[6]arene *pinched cone* conformation (Fig. 1.5, left) as + -, - + , + -, + -, - +, + - or C_1 (+ -, - +, + -)$_2$, although is also indicated as C_2 (+ -, - +, + -).

1.3. References

1. Gutsche C.D., *Calixarenes* (1989); *Calixarenes Revisited* in "Monograph in Supramolecular Chemistry", ed. Stoddart J. F. (The Royal Society of Chemistry, Cambridge, 1998).
2. Arduini A., Casnati A. (1996) "Calixarenes" in *Macrocyclic Synthesis: a Practical Approach*, ed. Parker D. (Oxford University Press, Oxford, 1996), 145–173. Pochini A., Ungaro R. in *Comprehensive Supramolecular Chemistry*, Vol. 2, Vögtle F. ed. (Pergamon Press, Oxford, 1996), 103–142. Böhmer, V. *Angew. Chem., Int. Ed. Engl.* **34** (1995), 713–745. Vicens J., Böhmer V. eds. *"Calixarenes: A Versatile Class of Macrocyclic Compounds"* (Kluwer Academic Publishers, Dordrecht, 1991).
3. Zinke A., Ziegler E., *Ber.* **74B** (1941), 205–214; 1729–1736.
4. Cornforth J.W., D'arcy Hart P., Nichollis G.A., Stock J.A., *Brit. J. Pharmacol.* **10** (1955), 73–86. Cornforth J.W., Morgan E.D., Potts K.T., Reees R.J.W., *Tetrahedron,* **29** (1973), 1659–1667.
5. For an account of the early history of calixarene chemistry see: Kappe T., *J. Incl. Phenom. Mol. Recogn. Chem.* **19** (1994), 3–15.
6. Gutsche C.D., Muthukrishan R., *J. Org. Chem.* **43** (1978), 4905–4906.
7. Andreetti G.D., Ungaro R., Pochini A., *J. Chem. Soc., Chem. Commun.,* (1979), 1005–1006.
8. For a recent review article on calixresorcarenes see: Timmerman P., Verboom W., Reinhoudt D.N., *Tetrahedron* **52** (1996), 2663–2704.
9. Masci B., Finelli M., Varrone M., *Chem. Eur. J.* **4** (1998), 2018–2030 and refences therein.
10. Gutsche C.D., Iqbal M., *Org. Synth., Coll. Vol. VIII*, (1993), 75–76.

11. Gutsche C. D., Dhawan B., Leonis M., Stewart D., *Org. Synth. Coll. Vol. VIII,* (1993), 77–78.
12. Munch J.H., Gutsche C.D., *Org. Synth., Coll. Vol. VIII*, (1993), 80–81.
13. Casnati A., Ferdani R., Pochini A., Ungaro R., *J. Org. Chem.* **62** (1997), 6236–624.
14. Lubitov I.E., Shokova E.A., Kovalev V.V., *Synlett* (1993), 647–648.
15. Kayes B.T., Hunter R.F., *J. Appl. Chem.* **8** (1958), 743–748.
16. Kämmerer H., Happel G., Caesar F., *Makromol. Chem.* **162** (1972), 179–197. Kämmerer H., Happel G., Mathiash W.B. *ibid.* **182** (1981), 1685–1694.
17. Böhmer V., Chhim P., Kämmerer H., *Makromol. Chem.* **180** (1979), 2503–2506. Böhmer V., Marschollek F., Zetta L., *J. Org. Chem.* **52**, (1987), 3200–3205.
18. Böhmer V., Merkel L., Kunz U., *J. Chem. Soc., Chem. Commun.* (1987), 896–897.
19. Alfieri C., Dradi E., Pochini A., Ungaro R., *Gazz. Chim.It.* **119** (1989), 335–338.
20. Shinkai S., *Tetrahedron* **49** (1993), 8933–8968 and references therein.
21. Jaime C, de Mendoza J., Prados P., Nieto P.M., Sanchez C., *J. Org. Chem.* **56** (1991), 3372–3376.
22. Gutsche C.D., Bauer L.J. *Tetrahedron Lett.* **22** (1981), 4763–4766.
23. Stewart D.R., Krawiec R.P., Kashyop R.P., Watson W.H., Gutsche C.D., *J. Am. Chem. Soc.* **117** (1995), 586–601.
24. Oksuka H., Shinkai S., *Supramol. Sci.* **31** (1996), 185–205 and references therein.
25. Neri P., Consoli G.M.L., Cunsolo F., Geraci C., Piattelli M., *New J. Chem.* **20** (1996), 433–446.
26. Andreetti G.D., Ugozzoli F., Casnati A., Ghidini, E., Pochini A., Ungaro R. *Gazz. Chim. Ital.* **119** (1989), 47–50.
27. Gutsche C.D., Gutsche A., Karaulov, A.I., *J. Incl. Phenom.* **3** (1985), 447–451.
28. Kanamathareddy S., Gutsche C.D., *J. Am. Chem. Soc.* **115** (1993), 6572–6579.
29. Ugozzoli F., Andreetti G.D., *J. Incl. Phenom. Mol. Recogn.* **13** (1992), 337–348.

CHAPTER 2

MOLECULAR MODELING OF CALIXARENES AND THEIR HOST-GUEST COMPLEXES

FRANK C. J. M. VAN VEGGEL

University of Twente, Faculty of Chemical Technology, Laboratories of Supramolecular Chemistry and Technology and MESA$^+$ research institute, P.O. Box 217, 7500 AE Enschede, The Netherlands.
Email: f.c.j.m.vanVeggel@ct.utwente.nl

2.1. Introduction

Simulations on the molecular level can be done with a variety of approaches, like ab-initio, semi-empirical, and empirical calculations. Only the latter have been used to study calixarenes, with a few exceptions. The empirical approaches have a classical description of molecules and are also called force field calculations. Well-known generic force fields are CHARMM,[1] AMBER,[2] BOSS,[3] MM2-4,[4] and TRIPOS.[5] Typical components of a force field are harmonic potentials for bonds, angles, and improper torsions, a cosine series for torsions, for the bonded interactions, and a Lennard-Jones potential and a coulombic term for the non-bonded interactions.[6] These terms are accompanied by a large set of parameters, completing the force field. So, bonds between two atoms, i.e. atom types, use two parameters to describe the harmonic potential; the force constant and the "natural" bond length. Similarly for angles, etc.

Some common steps in the process of performing these type of calculations are the following. Firstly, a starting structure has to be generated. Modern simulations packages offer 2D and 3D drawing facilities, but also allow to read in the coordinates of X-ray structures in a variety of formats. As an alternative, NMR data like distances between atom pairs, can be used as input. The next step is then the atom typing and the calculation of the atomic point charges. The description of the electrostatic interactions through point charges localized on the atomic coordinates is usually the weakest part of the force field. Charge redistributions in conformationally flexible molecules and polarization effects are not incorporated in most of the generic force fields. Nevertheless, it can be stated that the accuracy of modern force fields is quite high and that hence much insight can be obtained on the molecular level.

After the atom typing and calculations of the point charges the energy of the systems can be minimized, thus finding a local minimum on the potential energy surface. This technique is abbreviated as MM. Among the different algorithms that can be used to minimize the energy are steepest descent, conjugate gradient, adopted-basis set Newton Raphson, and Newton Raphson. Steepest descent is good to remove some of the worst interactions, but converges very slowly near the minimum. In this respect conjugate gradient and adopted-basis set Newton Raphson are much better. Only Newton Raphson uses second derivatives and is very efficient close to a minimum. Minimizations are stopped when a certain gradient is reached, e.g. 0.001 kcal/mol Å^2. It has to be realized that these algorithms can only go down in energy on the potential energy surface and hence "the nearest" local minimum will be generated. Methods to probe the entire potential energy surface in search for all relevant minima, including the global minimum, can be based on random moves followed by minimization, quenched dynamics, etc.[6]

If two (local) minima are connected via a conformational transition, it is in principle possible to calculated the energy barrier and the transition state. The transition state is a saddle point on the potential energy surface. Since in particular calix[4]arenes with substituents smaller than ethyl have conformational flexibility at ambient temperature, it is no wonder that studies as to that have appeared (see paragraph 2.4 for a discussion).

The structure that is the result of an MM calculations has no kinetic energy, thus it is a structure at 0 K. With molecular dynamics (MD) a more realistic picture can be obtained about the conformational behavior of the systems. In an MD simulations the nuclear motion of the atoms is calculated by solving numerically Newton's laws of motion; the Leapfrog Verlet algorithm provides one way to do this.[6] MD simulations can be done in the gas phase in order to probe the intrinsic flexibility of the system, but more often it is done in an explicit solvent model. An explicit solvent model describes a solvent in terms of discrete molecules and therefore also relies on a force field. An infinite solution can be simulated by making copies of the so-called primary box in the ±x, ±y, ±z directions, making a total of 27. In principle one would like to calculate the non-bonded interactions until infinity, but this is not practical and should also be avoided for technical reasons.[6] Via an MD simulation an ensemble, i.e. a set of conformations, is created that in principle describes the Boltzmann distribution and can be used to calculate properties like diffusion coefficients, etc.

An alternative way to generate an ensemble is via the Monte Carlo (MC) method.[7] The Metropolis algorithm is the most well-known for MC simulations. In this method configurations are generated by random moves of e.g. a solvent molecules, a torsion, etc. followed by an energy evaluation. If the energy is lower than the previous accepted random move, the configuration is always accepted. If the energy is higher, it is accepted by throwing a die and accepting it with a Boltzmann weighted chance. This is a very efficient method to generate an ensemble,[8] but one has to realize that there is no kinetic energy in this algorithm and properties like diffusion coefficients can thus not be calculated.

With the advent of fast computers, binding free energies (ΔG) of complexation became feasible. In principle one would like to calculate e.g. the complexation process of a receptor and a guest (see Figure 2.1). This is however, even with today's CPU, not possible. It is however possible to calculate this binding free energy change via a thermodynamic cycle. Since the Gibbs free energy is a state function, the closed path integral is by definition zero and hence the two horizontal *physical* processes can be related to the vertical *non-physical* processes. These latter two are feasible in a computer. The calculation of absolute and relative binding free energies, through the thermodynamic cycle, rely on free energy perturbation (FEP). In short, free energies changes can only be calculated accurately between two states if they are not too different. This means that e.g. going from a Na^+ to a K^+ complex is a too big step and therefore non-physical intermediate states are generated through a coupling parameter λ. For more details see the references cited.[6]

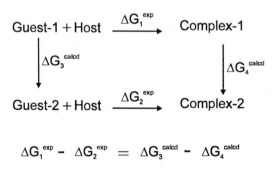

Figure 2.1 Thermodynamic Cycle

A search in the Chemical Abstracts[9] until January 1999 gave 159 articles. From this set roughly 100 articles were discarded for the following reasons. This review focuses on the calculation of the Boltzmann distribution, conformational transitions, and the complexation of neutral molecules and charged guests, eliminating a number of papers. Often, MM was only used as an "advanced CPK model" to support observed conformations, with a very scarce description of the applied method or none at all. This makes evaluation and reproduction of the results impossible, making it a rather useless scientific activity. Such articles were not considered.

It is assumed that the reader is familiar with the nomenclature of the extreme conformations of calix[n]arenes, i.e. cone, partial cone, 1,2-alternate, and 1,3-alternate for calix[4]arene, etc.[10]

2.2. Boltzmann Distribution

The Boltzmann distribution can in principle be calculated, if all relevant minima are included. In general, however, the agreement with experimental observations, often obtained in solution by NMR spectroscopy, is poor if a number of conformations is within an energy window of only a few kcal/mol. This can, however, be improved a lot if the vibrational and rotational contributions are included. The vibrational and rotational contributions can be calculated with equations 2.1 and 2.2, derived from statistical thermodynamics.[11] The Eigen frequencies can be calculated by a normal mode analysis of a well-minimized structure.

$$\Delta A_{vib} = -NkT \sum_{i=1}^{3n-6} [\frac{h\nu_i}{2kT} + \ln(1-e^{-h\nu_i/kT})] \qquad (2.1)$$

$$\Delta A_{rot} = -NkT\ln\left[\frac{\pi^{1/2}}{\sigma}\left[\frac{T^3}{\Theta_A\Theta_B\Theta_C}\right]^{1/2}\right] \qquad (2.2)$$

$$\Theta_x = \frac{h^2}{8\pi^2 I_x kT}$$

With: I_x = Moment of Inertia

σ = Rotational symmetry number

2.2.1. Calix[4]arenes

Due to limited computer power, the earliest reports on simulations on calixarenes were necessarily restricted to gas-phase minimizations. Calix[4]arenes with four hydroxyl groups on the lower rim are stabilized in the cone conformation by a circular array of four hydrogen bonds, making it the most stable conformation and the only one detectable by NMR. Force field calculations have no trouble in reproducing this.[12,13,14,15,16,17] The *para-* and *meta*-substituents have little or no influence on the relative distribution of the four conformations. Lipkowitz and Pearl[18] reported MM studies on the native calix[4]arene and its tetramethyl ether with MM2, AMBER, OPLSA, CHARMm, MOPAC6 AM1 and PM3, and AMPAC AM1 and PM3 and concluded these methods quite inappropriate. In particular CHARMm performed badly, but van Hoorn et al.[19] showed that their results were erroneous.

The tetramethyl calix[4]arene has a conformationally very rich behavior and the Boltzmann distribution is strongly dependent on, among others, the solvent. In chloroform solution the Boltzmann distribution of cone : partial cone : 1,2-alternate : 1,3-alternate is 4 : 85 : 8 : 3.[20] In dichloromethane the partial cone is still the most abundant conformation, but more cone is present. The distribution is a strong function of the temperature (190 - 255 K) with the partial cone in 60 - 75 %, the cone in 40-20 %, and the other two in only a few percent present. Many minimizations using various force fields gave widely different results, sometimes not even predicting the partial cone to be present.[20,21] If however all in/out position of the methoxy groups are systematically taken into account and corrections for the vibrational and rotational contributions are made with the equations from statistical thermodynamics, the Boltzmann distribution is in good agreement with that in chloroform[20] (calculated: 2 : 82 : 5 : 11). Agreement of this gas-phase calculated Boltzmann distribution with that in dichloromethane is very poor, but the more abundant presence of the cone conformation could be explained by an inclusion of one molecule of dichloromethane in the cavity. Such favorable inclusion is not present in chloroform.[19] These results were later corroborated by Hirsch et al.[22]

Two theoretical studies were reported on the partially methylated calix[4]arenes,[17,23] the latter showing a good agreement between the calculated Boltzmann distribution, including vibrational and rotational contributions, and the NMR determined distributions [in various (chlorinated) solvents].

Three groups also studied the effect of OH-depleted calix[4]arenes.[14a,15,16] Little differences between the force fields [CHARMM and twice MM3(92)] were observed, with the energy differences between the conformations being much smaller than in the case of the native calix[4]arene.

2.2.2. Calix[5]arenes

The number of theoretical studies on calix[5]arenes is quite limited,[24] probably due to the fact that it is synthetically less accessible. The cone conformation of *para*-methylcalix[5]arene was reproduced by MM3, but not by TRIPOS calculations. The TRIPOS force field did however, reproduce the cone of calix[5]arenes having an equatorial substituent at one of the methylene bridges. The MM3 predicted stabilities of the pentamethyl ether of *para*-methylcalix[5]arene was 1,3-alternate >

paco > 1,2-alternate > cone, whereas with TRIPOS the order was 1,2-alternate > paco > 1,3-alternate > cone. Energy differences are however small, suggesting that various conformations exist, which has indeed experimental precedent.

2.2.3. Calix[6]arenes

Van Hoorn et al.[25] calculated with CHARMM that the *pinched* cone conformation of hexahydroxycalix[6]arene was in better agreement with the experimental evidence than the *winged* cone. The *winged* cone conformation had previously been reported as a lower energy conformations by de Mendoza[26] and they recently refuted the reinterpretation by van Hoorn et al.[27]

The conformational behavior of the 1,2-*para-tert*-butylbenzyl ether of *para-tert*-butylcalix[6]arene has been studied (MM2) by Neri et al.[28] They generated 21 possible conformations with different "up-down" orientations of the phenolic rings, followed by minimization. The minimized structures could be grouped in two classes, i.e. a *syn* and an *anti* orientation of the two ether substituents, in agreement with isolated compounds. In one class there is a fast exchange between conformations involving rotations of hydroxyl-bearing rings.

Similarly, the conformational behavior of the 1,2,4,5-tetra(2-pyridylmethyl) ether of *para-tert*-butylcalix[6]arene was studied. 14 Conformations were calculated giving the *anti*-1,2-*anti*-4,5 conformation as the most stable (MM2), in agreement with NMR data.[29]

The conformational properties of 1,3,5-trimethoxycalix[6]arene with various substituents at the *para*-positions have been investigated by van Hoorn et al.[30] All possible orientations of the aromatic rings (up and down) and of the *para*-substituent (pointing outward or inward) were considered, showing that the cone with the three substituents pointing outward is the most stable conformation with the three methoxy groups filling the cavity. This is in good agreement with experimental findings.

2.3. Complexes with neutral molecules

Gutsche and See[31] studied the complexation of "double-cavity calix[4]arenes (see Chart) with respect to phenols, carboxylic acids, pyridines, imidazoles, and aliphatic amines and reported association constants in the order of <5 to 55 M^{-1}. They rationalized their results by Quanta/CHARMm minimizations, which suggested a side on binding rather than binding through/in the bottom cavity.

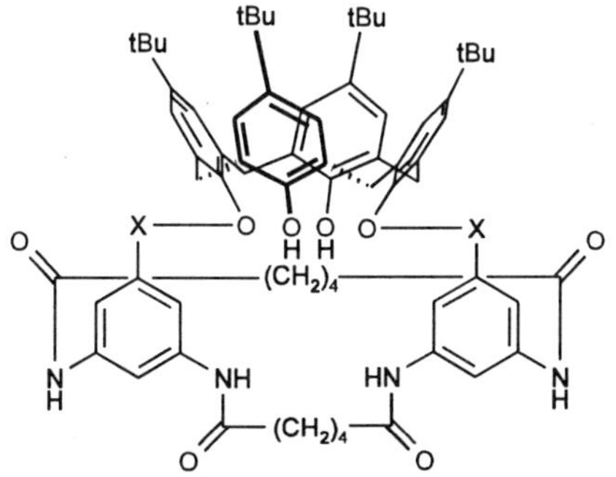

The stable inclusion of dichloromethane, as opposed to chloroform, in the cone conformation of tetramethoxy-*para-tert*-butylcalix[4]arene was inferred from MD simulations in explicit solvent models and gave a rational of the increased population of the cone conformation in dichloromethane compared to chloroform.[19]

The gas phase interaction energies between vapors like perchloroethylene, chloroform, dichloromethane, tetra, benzene, toluene, and xylene and calixarenes and alike have been successfully correlated with sensor responses of quartz crystal microbalances and surface acoustic wave devices.[32,33]

The 2:1 solid state complex of *para-tert*-butylcalix[4]arene and *para*-xylene has been investigated by inelastic neutron scattering (INS).[34] From the MM3 minimized structure, the normal-mode frequencies and atomic displacements were calculated, which provided the required input for the INS data analysis. These calculations also gave partial evidence of the dynamics of the system, suggesting a rotation around the long axis of the *para*-xylene.

2.4. Conformational interconversions

The conformational flexibility of calixarenes has been studied both experimentally, mostly by NMR spectroscopy, and theoretically. The aim of theoretical studies is to identify transition states, i.e. saddle points on the potential energy surface, connecting two (local) energy minima. The energy difference $\Delta E^{\neq}$ between minimum and transition state is in general a good approximation to the $\Delta H^{\neq}$ of the process and sometimes even $\Delta G^{\neq}$ if the $\Delta S^{\neq}$ is small. This is often the case for conformational transitions. A true transition state should have one negative Eigen frequency, the reaction coordinate, that can be calculated by a normal mode analysis. We believe that such a proof is important, because not all used algorithms guarantee that indeed a saddle point has been calculated.

The parent calix[4]arene shows a rapid interconversion from the cone to the inverted cone in chloroform at room temperature ($\Delta G^{\neq} \approx 15$ kcal/mol; $\Delta H^{\neq}$ 14.2 kcal/mol).[35] It has been debated for quite some time what the mechanism of this process was. Two mechanisms have been proposed: a "continuous-chain" mechanism in which the circular array of four hydrogen bonds is not broken and a "broken-chain" in which one phenolic rings flips through the annulus, followed by the others. In this latter mechanism an opposite or adjacent phenolic ring could flip after the first ring flip.

2.4.1. Tetrahydroxycalix[4]arenes

One of the earliest reports on this matter was by Grootenhuis et al.[13] They performed MD simulations (AMBER) in the gas phase and in water and concluded that the partial cone was the key intermediate in the interconversions. MMP2 calculations on *para-iso*-propylcalix[4]arene with a constrained torsion angle passing through the methylene groups of a phenolic ring showed an activation energy of 8.14 kcal/mol for the cone to partial cone interconversion and smaller barriers for the partial cone to 1,2- and 1,3-alternate.[36] Fischer et al.[15] used the "conjugate peak refinement" algorithm, implemented as the TRAVEL module in CHARMM, that searches for true saddle points on the potential energy surface. As input only two well-minimized minima are required without any further constraint. The rate-limiting step in the interconversion from cone to inverted cone is the partial cone ($\Delta E^{\neq}$ = 14.5 kcal/mol). The next step is predicted to be the transition to the 1,2-alternate (Figure 2.2).

Thondorf et al.[16,37] used an approach referred to as the "coordinate driver method" to identify transition states connecting two minima. In later reports the saddle points were actually checked by a normal mode analysis to have one negative Eigen value. They found barriers in the order of 11 kcal/mol for the cone to partial cone interconversion and in the order of 14 kcal/mol for the partial cone to 1,2-alternate. Transitions from the partial cone to the 1,3-alternate have somewhat higher barriers. Since the partial cone is not observed by NMR spectroscopy, the predicted rate-limiting step is the conversion from the partial cone to 1,2-alternate with a $\Delta E^{\neq}$ of 13.7 kcal/mol. This prediction is not in agreement with those from Fischer et al., in contrast to what the authors concluded. Now, there is considerable evidence that the cone to inverted cone proceeds in a stepwise manner, with the partial cone playing a key role.

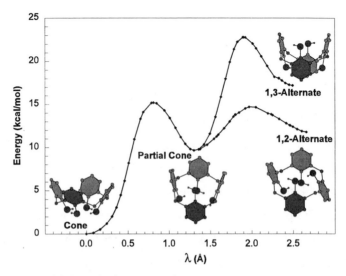

Figure 2.2 Stepwise interconversion

The circular array of four hydrogen bonds has either a clock wise or counter clock wise arrangement, preventing the calix to adopt "pure" C_{2v} or C_{4v} symmetry. The quasi C_{4v} symmetry is usually not observed by NMR spectroscopy, but is the transition state between the two pinched cone conformations with quasi C_{2v} symmetry. Barriers for this interconversion have been calculated to be 3.1 kcal/mol[15] and 0.6 kcal/mol[14b], consistent with a rapid equilibrium at ambient conditions giving an apparent quasi C_{4v} symmetry.

Den Otter and Briels[38] were the first to actually calculate the rate of interconversion of the de-*tert*-butylated calix[4]arene, in the gas phase as well as in explicit solvent models, using the so-called reactive flux method. They used the unstable normal mode of the transition state in MD simulations to calculate this rate and found very good agreement with experimental results. Even subtle solvent effects were reproduced with good accuracy.

Due to the fact that the cone is stabilized by four hydrogen bonds, it is not surprising that OH depleted calixarenes show much smaller barriers for interconversions.[14b,15,16]

2.4.2. (Partially) O-alkylated calix[4]arenes

The energy barriers between the key partial cone conformation of tetramethoxycalix[4]arene and the cone, 1,2-, and 1,3-alternate were calculated to be 19.6, 20.2, and 18.2 kcal/mol, respectively; in qualitative agreement with relative rates deduced from 2D EXSY NMR.[15] The in/out rotation of a methoxy group has an energy barrier of only 6-8 kcal/mol, consistent with the upper bound value obtained from the NMR time scale. Tetramethyl and tetra-*n*-propyl ethers of calix[4]arenes, with Br, NO_2, NH_2, and *tert*-Bu groups at the *para*-positions, have been studied by Hirsch et al.[22] The PM3 calculated barriers for the pinched cone/pinched cone interconversion were calculated to be only 0.8 to 1.6 kcal/mol, with a C_{4v} symmetric transition state.

The conformational interconversions from the cone to the inverted cone of the monomethyl, 1,2-dimethyl, 1,3-dimethyl, and trimethyl ethers of calix[4]arene have been systematically investigated by van Hoorn et al.[23] and the calculated energy barriers are in qualitative agreement with experimental results. Compared to the parent calix[4]arene and its tetramethyl ether the barriers are much higher, which is a cooperative effect of hydrogen bonding and steric crowding.

2.4.3. Calix[5]- and calix[6]arenes

The conformational behavior of penta-*para*-methylcalix[5]arene has been studied by MD simulations in the gas phase[24a] showing both pseudorotation of methylene groups and ring flips of phenolic rings. A TRAVEL analysis on the possible interconversions of hexahydroxycalix[6]arene showed that experimentally observed pseudorotation and inversion are likely one concerted pseudorotation/inversion process of the pinched cone.[25] As stated above, de Mendoza and coworkers arrived at a different conclusion.[27]

It is well-known that it is much more difficult to fix calix[6]arene in a certain conformation by derivatizing the lower rim, because also the upper rim part is able to rotate through the annulus. Van Hoorn et al.[30] reproduced the experimentally found upper limit for interconversion by TRAVEL calculations on the 2,4,6-substituted 1,3,5-trimethoxycalix[6]arene with *para-tert*-butyl groups. If the 2,4,6-substituents become too large to rotate through the annulus, the *tert*-butyl group "take over" the interconversion with a calculated upper limit of 17.5 kcal/mol, reasonably close to the experimental value of approximately 21 kcal/mol.

2.5. Complexes of monovalent cations and anions

Theoretical studies on the binding affinities and selectivities of calixarenes have been focused to a large extent on the Na^+ versus K^+ and on Na^+ versus Cs^+. The latter topic has been inspired by the extraction of radioactive Cs^+ from waste stream containing high concentrations of $NaNO_3$ and HNO_3, as a result of the recovery of uranium and plutonium from nuclear fuel bars.

Reinhoudt and coworkers developed calixspherands (see Chart) for the kinetically inert complexation of alkali metal ions (binding free energies in the order of -15 kcal/mol).[39] Miyamoto and Kollman[40] calculated relative and absolute free energies of binding in water which were in very good agreement with experiments.

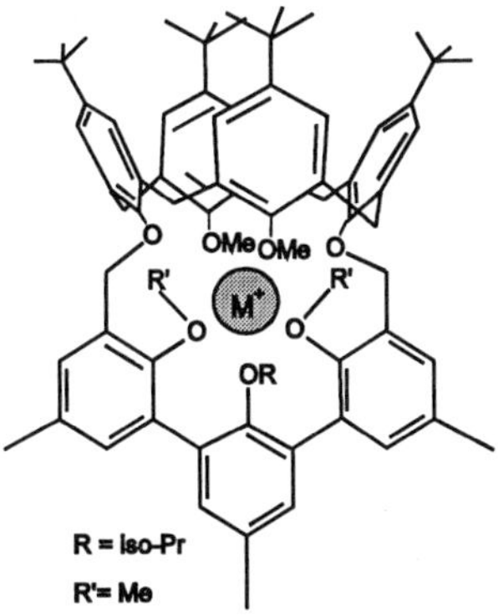

R = Iso-Pr
R'= Me

The lower rim tetraesters and tetraamides (**1** and **2**) exhibit a high selectivity of Na^+ over K^+.[41]

Wipff et al.[42] calculated relative free energies in the gas phase, water, and acetonitrile for the tetraamide derivative. They found an intrinsic selectivity for Li^+ over the other alkali cations, but in water no definite conclusions could be drawn. The simulations in acetonitrile showed a qualitative agreement with available experimental work on the tetraester derivative. Lower rim ketones of calix[5]- and calix[6]arenes have been studied by Bell et al. (**3**).[43] Their MM calculations are quite similar to the X-ray analyses, but otherwise the theoretical part is rather non-conclusive.

1, R1 = R2 = CH$_2$C(O)NEt$_2$; n = 3
2, R1 = R2 = CH$_2$C(O)OAlkyl ; n = 3
3, R1 = R2 = CH$_2$C(O)Alkyl ; n = 4,5
5, R1 = R2 = CH$_2$COO$^-$; R3 = H ; n = 3
6, R1 = R2 = CH$_2$COO$^-$; R3 = SO$_3^-$; n = 3
7, R1 = CH$_2$C(O)NHR4; R4 = n-Pr, (CH$_2$)$_n$-2-Py
R2 = CH$_2$COO$^-$; n = 3

8, R1 = R2 = [bipyridyl group]   R3 = t-Bu
n = 3

9, R1 = R2 = [bipyridyl group]   R3 = H
n = 3

10, R1 = [bipyridyl group]   R2 = n-Pr
n = 3   R3 = t-Bu

Calix[4]crowns (e.g. **4**) have been developed for the separation of small amounts of Cs$^+$ from Na$^+$ rich waste streams.

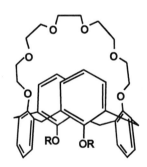

4 1,3-alternate-calix[4]-crown-6

Wipff and coworkers[44] have performed detailed theoretical analyses, including free energy perturbation calculations, on the Cs^+ over Na^+ selectivity. In general the 1,3-alternate conformation is selective for Cs^+, but solvent effects can have a large influence. Lamare and coworkers[45] have performed MD simulations in the gas phase and in water, providing evidence for the lack of affinity for Na^+ of the mono- and biscrown-6 1,3-alternate calix[4]arene and they also concluded that the selectivity of Cs^+ over Na^+ is probably more due to a mismatch for Na^+ than to a reorganization for Cs^+ and/or the worse solvation of Na^+ in a rather hydrophobic cavity. Free energy perturbation calculations in water predicted a Cs^+ over Na^+ selectivity, qualitatively in agreement with data in methanol.

An MD simulation in chloroform of a 1,4-bridged calix[6]arene-crown-5, with amide groups on the four remaining phenol rings, showed no preferred binding mode for $Me_4N^+OAc^-$ and an exo-mode for $NH_4^+OAc^-$.[46] MM calculations on the binding mode of the tetracarboxylate calix[4]arene and tetracarboxylate-(tetrasulfonate) calix[4]arene (**5** and **6**) with *N,N,N*-trimethylanilinium, benzyltrimethylammonium, and *para*-nitrobenzyltrimethylammonium supported the NMR-deduced deeper binding of the guest in the host **6**.[47]

The halide binding of the bisurea host, developed by Reinhoudt c.s.,[48] was theoretically investigated by Jorgensen and coworkers (see Chart).[49]

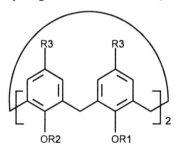

R1 = n-Pr; R3 = tert-Bu
R2 = $(CH_2)_4NHC(O)NHPh$

The relative binding free energy differences of Cl⁻, Br⁻, and I⁻ could accurately be reproduced in (dry) chloroform, but the calculations failed to reproduce the non-binding of F⁻. This was subsequently explained by including some water molecules in the calculations, which reversed the intrinsic selectivity of F⁻ over the other halide ions.

2.6. Complexes of divalent cations

Compared to simulations on monovalent cations, little work has been done on divalent cations. The calculated free energy differences of the tetraamide calix[4]arene (**1**), with very similar results for water and methanol, reproduced the experimental data in methanol, namely $Ca^{2+} > Sr^{2+} > Ba^{2+} \gg Mg^{2+}$.[50] In this paper also the uranyl cation (UO_2^{2+}) was investigated, being bound less strongly than the alkaline-earth cations. In two other papers by Wipff and coworkers[51] the selective binding in aqueous solution of the uranyl cation by the *para*-methylcalix[6]arene⁶⁻ ion over Sr^{2+} was investigated. Harriman et al.[52] studied the calix[4]diquinones possessing luminescent Ru^{2+}-trisbipyridyl moieties. Without binding of another cation the photoexcited luminescence of the ruthenium complex is quenched by the quinone moieties, which was rationalized by MD simulations showing van der Waals contact between these two parts. Upon binding of cations such as Ba^{2+} this quenching is reduced, which was explained by a more remote position of luminescent center and quencher upon metal binding. This explanation was corroborated by MD simulations.

2.7. Complexes of trivalent cations

The required efficient shielding of luminescent lanthanide ions such as Eu^{3+} and Tb^{3+} has inspired the use of calix[4]arene-derived ligands.

The Eu^{3+} complex of the tetraamide calix[4]arene (**1**) has been theoretically investigated in vacuo, water, and acetonitrile.[42] As model for Eu^{3+} the authors used the Lennard-Jones parameters of Na^+ with a three plus charge. The rational for this approximate model was that Na^+ and Eu^{3+} have about the same ionic radius. In the gas phase the Eu^{3+} is bound in the cavity provided by the four phenolic and four amide oxygens, but an MD in explicit water showed a strong competition with water. The Eu^{3+} ion became only bound to the four amide oxygens with room for four waters. In acetonitrile the Eu^{3+} is more deeply bound in the host compared to water, with one acetonitrile in the first coordination sphere. Van Veggel and Reinhoudt[53] performed MD simulations in an explicit methanol model (OPLS) on the Eu^{3+} complexes of calix[4]arene-derived ligands with a potential ninth coordination site for improved solvent shielding (**7**). As Eu^{3+} model they used a Lennard-Jones parameter set with a three plus charge that reproduced the coordination number in methanol. It was concluded that the ninth coordination site was not beneficial for the shielding and that up to three molecules of methanol are in the first coordination sphere, all in good agreement with experimental data. Similar general conclusions were drawn by Grote Gansey et al.[54] in a study on the binding of Ac^{3+} by calix[4]arene-derived ligands having deprotonable ninth coordination sites. The model used for Ac^{3+} reproduced the hydration free energy within 1% of the experiment.[55] Van Veggel[56] reported MM, MD, and FEP-MC calculations on the Eu^{3+} complexes of calix[4]arene ligands having 3 or 4 bipyridyl moieties (**8-10**). The experimental finding that ligand **9** does not give stable complexes, whereas the other two do, could be rationalized by MD simulations in explicit acetonitrile (OPLS). The calculations including nitrate counter ions showed that they can not be neglected in non-protic solvents. Similar findings were reported by Ulrich et al.[57]

2.8. Simulations at the water-organic solvent interface

In an effort to get a better understanding of the behavior of ligands extracting cations at the water-organic solvent interface, Wipff et al.[58] have performed a number of theoretical studies. They concluded that ligands based on calixarenes behave like surfactants and that counter ions, lipophilic versus hydrophilic, have a large influence. The selectivity of extraction of Cs^+ over Na^+ could qualitatively be reproduced by an elaborated thermodynamic cycle of the binary system.

2.9. General conclusions

This review shows that modern simulation techniques like molecular mechanics and dynamics and free energy perturbation simulations are capable of reproducing experimental observation, qualitatively and even quantitatively. Boltzmann distributions, specific solvent inclusion/effects, energy barriers and pathways of interconversions, absolute and relative binding free energies of cations and anions, solvent shielding of trivalent cations are topics where these techniques can successfully be applied.

2.10. References

[1] (a) Brooks, B. R.; Bruccoleri, R. E.; Olafsen, B. D.; States, D. J.; Swaminathan, S.; Karplus, M. *J. Comput. Chem.* **4** (1983), 187-217; (b) Momany, F. A.; Klimkowski, V. J.; Schäfer, L. *J. Comput. Chem.* **11** (1990), 654-662; (c) Momany, F. A.; Rone, R.; Kunz, H.; Frey, R. F.; Newton, S. Q.; Schäfer, L. *J. Mol. Structure* **286** (1993), 1-18.

[2] (a) Kollman, P. A. *Chem. Rev.* **93** (1993), 2395-2417; (b) Kollman, P. A. *Acc. Chem. Res.* **29** (1996), 461-469; (c) Weiner, S. J.; Kollman, P. A.; Nguyen, D. T.; Case, D. A. *J. Comput. Chem.* **7** (1986), 230-252.

[3] (a) Jorgensen, W. L. *Acc. Chem. Res.* **22** (1989), 184-189; (b) Jorgensen, W. L.; Maxwell, D. S.; Tirado-Rives, J. *J. Am. Chem. Soc.* **118** (1996), 11225-11236; (c) Damm, W.; Frontera, A.; Tirado-Rives, J.; Jorgensen, W. L. *J. Comput. Chem.* **16** (1997), 1955-1970.

[4] (a) Hay, B. P.; Rustad, J. R. *J. Am. Chem. Soc.* **116** (1994), 6316-6326; (b) Allinger, N. L.; Chen, K.; Lii, J.-H. *J. Comput. Chem.* **17** (1996), 642-668; (c) Nevins, N.; Chen, K.; Allinger, N. L. *J. Comput. Chem.* **17** (1996), 669-694; (d) Nevins, N.; Lii, J.-H.; Allinger, N. L. *J. Comput. Chem.* **17** (1996), 695-729; (e) Nevins, N.; Allinger, N. L. *J. Comput. Chem.* **17** (1996), 730-746; (f) Allinger, N. L.; Chen, K.; Katzenellenbogen, J. A.; Wilson, S. R.; Anstead, G. M. *J. Comput. Chem.* **17** (1996), 747-755.

[5] Clark, M.; Cramer, III, R. D.; Opdenbosch, N. *J. Comput. Chem.* **10** (1989), 982-1012.

[6] (a) Lipkowitz, K. B.; Boyd, D. B. (Eds) *Reviews in Computational Chemistry* (VCH, Weinheim, 1990) vol 1-10; (b) Von Ragué Schleyer (Editor in Chief) *Encyclopedia of Computational Chemistry* (John Wiley & Sons, New York, 1998) vol 1-5. For a comparison of various force fields see for instance (a) Gundertofte, K.; Liljefors, T.; Norrby, P.; Pettersson, I. *J. Comput. Chem.* **17** (1996), 429-449; (b) Kaminski, G.; Jorgensen, W. L. *J. Phys. Chem.* **100** (1996), 18010-18013.

[7] Jorgensen, W. L. *J. Phys. Chem.* **87** (1983), 5304-5314.

[8] Jorgensen, W. L.; Tirado-Rives, J. *J. Phys. Chem.* **100** (1996), 14508-14513.

[9] The CA was searched on appropriately spelled MM, MD, MC, FF, various mentioned acronyms of FFs, and calix(?), giving 159 hits (January 21st, 1999).

[10] Gutsche, C. D. *Calixarenes Revisited* in *Monographs in Supramolecular Chemistry* Stoddart, J. F. (Series Editor) (The Royal Society of Chemistry, 1998).

[11] Hill, T. L.; *An Introduction to Statistical Thermodynamics* (Addison-Wesley Publishing Company, 1960).

[12] Bayard, F.; Decoret, C.; Pattou, D.; Royer, J.; Satrallah, A.; Vicens, J. *J. Chim. Phys.* **86** (1989), 945-954.

[13] Grootenhuis, P. D. J.; Kollman, P. A.; Groenen, L. C.; Reinhoudt, D. N.; van Hummel, G. J.; Ugozzoli, F.; Andreetti, G. D. *J. Am. Soc. Chem.* **112** (1990), 4165-4176.

[14] (a) Harada, T.; Rudzinski, J. M.; Osawa, E.; Shinkai, S. *Tetrahedron* **49** (1993), 5941-5954; (b) Harada, T.; Ohseto, F.; Shinkai, S. *Tetrahedron* **50** (1994), 13377-13394; (c) Harada, T.; Shinkai, S. *J. Chem. Soc., Perkin Trans. 2* (1995), 2231-2242.

[15] Fischer, S.; Grootenhuis, P. D. J.; Groenen, L. C.; van Hoorn, W. P.; van Veggel, F. C. J. M.; Reinhoudt, D. N.; Karplus, M. *J. Am. Soc. Chem.* **117** (1995), 1611-1620.

[16] Thondorf, I.; Brenn, J. *J. Mol. Struct. (Theochem)* **398-399** (1997), 307-314.

[17] Thondorf, I.; Hillig, G.; Brandt, W.; Brenn, J.; Barth, A.; Böhmer, V. *J. Chem. Soc., Perkin Trans. 2* (1994), 2259-2267.

[18] Lipkowitz, K. B.; Pearl, G. *J. Org. Chem.* **58** (1993), 6729-6736.

[19] van Hoorn, W. P.; Briels, W. J.; van Duynhoven, J. P. M.; van Veggel, F. C. J. M.; Reinhoudt, D. N. *J. Org. Chem.* **63** (1998), 1299-1308.

[20] Groenen, L. C.; van Loon, J.-D.; Verboom, W.; Harkema, S.; Casnati, A.; Ungaro, R.; Pochini, A.; Ugozzoli, F.; Reinhoudt, D. N. *J. Am. Soc. Chem.* **113** (1991), 2385-2392.

[21] A summmary given in ref 19.

[22] Soi, A.; Bauer, W.; Mauser, H.; Moll, C.; Hampel, F.; Hirsch, A. *J. Chem. Soc., Perkin Trans. 2* (1998), 1471-1478.

[23] van Hoorn, W. P.; Morshuis, M. G. H.; van Veggel, F. C. J. M.; Reinhoudt, D. N. *J. Phys. Chem. A* **102** (1998), 1130-1138.

[24] (a) Thondorf, I.; Brenn, J. *J. Chem. Soc., Perkin Trans. 2* (1997), 2293-2299; (b) Biali, S. E.; Böhmer, V.; Columbus, I.; Ferguson, G.; Grüttner, C.; Grynszpan, F.; Paulus, E. F.; Thondorf, I. *J. Chem. Soc., Perkin Trans. 2* (1998), 2261-2269.

[25] van Hoorn, W. P.; van Veggel, F. C. J. M.; Reinhoudt, D. N. *J. Org. Chem.* **61** (1996), 7180-7184.

[26] Molins, M. A.; Nieto, P. M.; Sánches, C.; Prados, P.; de Mendoza, J.; Pons, M. *J. Org. Chem.* **57** (1992), 6924-6931.

[27] Magrans, J. O.; Rincón, A. M.; Cuevas, F.; López-Prados, J.; Nieto, P. M.; Pons, M.; Prados, P.; de Mendoza, J. *J. Org. Chem.* **63** (1998), 1079-1985.

[28] Neri, P.; Rocco, C.; Consoli, G. M. L.; Piattelli, M. *J. Org. Chem.* **58** (1993), 6535-6537.

[29] Neri, P.; Foti, M.; Ferguson, G.; Gallagher, J. F.; Kaitner, B.; Pons, M.; Molins, M. A.; Giunta, L.; Pappalardo, S. *J. Am. Soc. Chem.* **114** (1992), 7814-7821.

[30] van Hoorn, W. P.; van Veggel, F. C. J. M.; Reinhoudt, D. N. *J. Phys. Chem. A* **102** (1998), 6676-6681.

[31] Gutsche, C. D.; See, K. A. *J. Org. Chem.* **57** (1992), 4527-4539.

[32] (a) Dominik, A.; Roth, H. J.; Schierbaum, K. D.; Göpel, W. *Supramol. Science* **1** (1994), 11-19; (b) Dickert, F. L.; Schuster, O. *Mikrochim. Acta* **119** (1995), 55-62; (c) Dickert, F. L.; Bäumler, U. P. A.; Stathopulos, H. *Anal. Chem.* **69** (1997), 1000-1005.

[33] For a review on mass sensors and supramolecular chemistry see: van Veggel, F. C. J. M. *Mass Sensors* in *Comprehensive Supramolecular Chemistry*, Editor in Chief Lehn, J.-M. (Pergamon, New York,1996), 171-185.

[34] Paci, B.; Deleuze, M. S.; Caciuffo, R.; Tomkinson, J.; Ugozzoli, F.; Zerbetto, F. *J. Phys. Chem. A* **102** (1998), 6910-6915.

[35] Araki, K.; Shinkai, S.; Matsuda, T. *Chem. Lett.* (1989), 581-584.

[36] Royer, J.; Bayard, F.; Decoret, C. *J. Chim. Phys.* **87** (1990), 1695-1700.

[37] (a) Thondorf, I.; Brenn, J.; Brandt, W.; Böhmer, V. *Tetrahedron. Lett.* **36** (1995), 6665-6668; (b) Biali, S. E.; Böhmer, V.; Brenn, J.; Frings, M.; Thondorf, I.; Vogt, W.; Wöhnert, J. *J. Org. Chem.* **62** (1997), 8350-8360; (c) Wöhnert, J.; Brenn, J.; Stoldt, M.; Aleksiuk, O.; Grynszpan, F.; Thondorf, I.; Biali, S. E. *J. Org. Chem.* **63** (1998), 3866-3874.

[38] (a) den Otter, W. K.; Briels, W. J. *J. Chem. Phys.* **106** (1997), 5494-5508; (b) den Otter, W. K.; Briels, W. J. *J. Chem. Phys.* **107** (1997), 4968-4978; (c) den Otter, W. K.; Briels, W. J. *J. Am. Chem. Soc.* **120** (1998), 13167-13175.

[39] Ghidini, E.; Ugozzoli, F.; Ungaro, R.; Harkema, S.; El-Fadl, A. A.; Reinhoudt, D. N. *J. Am. Chem. Soc.* **112** (1990), 6979-6985; and references cited therein.

[40] Miyamoto, S.; Kollman, P. A. *J. Am. Chem. Soc.* **114** (1992), 3668-3674.

[41] See ref. 10, pages 150-154.

[42] (a) Guilbaud, P.; Varnek, A.; Wipff, G. *J. Am. Chem. Soc.* **115** (1993), 8298-8312; (b) Varnek, A.; Wipff, G. *J. Phys. Chem.* **97** (1993), 10840-10848.

[43] Bell, S. E. J.; Browne, J. K.; McKee, V.; McKervey, M. A.; Malone, J. F.; O'Leary, M.; Walker, A. *J. Org. Chem.* **63** (1998), 489-501.

[44] (a) Wipff, G.; Lauterbach, M. *Supramol. Science* **6** (1995), 187-207; (b) Varnek, A.; Wipff, G. *J. Mol. Struct. (Theochem)* **363** (1996), 67-85.

[45] (a) Thuéry, P.; Nierlich, M.; Lamare, V.; Dozol, J.-F.; Asfari, Z.; Vicens, J. *Supramol. Science* **8** (1997), 319-332; (b) Thuéry, P.; Nierlich, M.; Bryan, J. C.; Lamare, V.; Dozol, J.-F.; Asfari, Z.; Vicens, J. *J. Chem. Soc., Dalton Trans.* (1997), 4191-4202; (c) Lamare, V.; Bressot, C.; Dozol, J.-F.; Vicens, J.; Asfari, Z.; Ungaro, R.; Casnati, A. *Separation Science Techn.* **32** (1997), 175-191; (d)

Lamare, V.; Dozol, J.-F.; Ugozzoli, F.; Casnati, A.; Ungaro, R. *Eur. J. Org. Chem.* (1998), 1559-1568; (e) Lamare, V.; Dozol, J.-F.; Fuangswasdi, S.; Arnaud-Neu, F.; Thuéry, P.; Nierlich, M.; Asfari, Z.; Vicens, J. *J. Chem. Soc., Perkin Trans. 2* (1999), 271-284.

[46] Fraternali, F.; Wipff, G. *Anales Química Int. Ed.* 93 (1997), 376-384.

[47] Arena, G.; Casnati, A.; Contino, A.; Lombardo, G. G.; Sciotto, D.; Ungaro, R. *Chem. Eur. J.* 5 (1999), 738-744.

[48] Scheerder, J.; Fochi, M.; Engbersen, J. F. J.; Reinhoudt, D. N. *J. Org. Chem.* 59 (1994), 7815-7820.

[49] McDonald, N. A.; Duffy, E. M.; Jorgensen, W. L. *J. Am. Chem. Soc.* 120 (1998), 5104-5111.

[50] Muzet, N.; Wipff, G.; Casnati, A.; Domiano, L.; Ungaro, R.; Ugozzoli, F. *J. Chem. Soc., Perkin Trans. 2* (1996), 1065-1075.

[51] (a) Guilbaud, P.; Wipff, G. *J. Mol. Struct. (Theochem)* 366 (1996), 55-63; (b) Guilbaud, P.; Wipff, G. *J. Inclusion Phenom. Molec. Recogn. Chem.* 16 (1993), 169-188.

[52] Harriman, A.; Hissler, M.; Jost, P.; Wipff, G.; Ziessel, R. *J. Am. Chem. Soc.* 121 (1999), 14-27.

[53] van Veggel, F. C. J. M.; Reinhoudt, D. N. *Recl. Trav. Chim. Pays-Bas* 114 (1995), 387-394.

[54] Grote Gansey, M. H. B.; Verboom, W.; van Veggel, F. C. J. M.; Vetrogon, V.; Arnaud-Nue, F.; Schwing-Weill, M.-J.; Reinhoudt, D. N. *J. Chem. Soc., Perkin Trans. 2* (1998), 2351-2360.

[55] For accurate Lennard-Jones parameters for Ln^{3+} see: van Veggel, F. C. J. M.; Reinhoudt, D. N. *Chem. Eur. J.* 5 (1999), 90-95.

[56] van Veggel, F. C. J. M. *J. Phys. Chem. A* 101 (1997), 2755-2765.

[57] Ulrich, G.; Ziessel, R.; Manet, I.; Guardigli, M.; Sabbatini, N.; Fraternali, F.; Wipff, G. *Chem. Eur. J.* **3** (1997), 1815-1822.

[58] (a) Wipff, G.; Engler, E.; Guilbaud, P.; Lauterbach, M.; Troxler, L.; Varnek, A. *New J. Chem.* **20** (1996), 403-417; (b) Varnek, A.; Wipff, G. *J. Comput. Chem.* **17** (1996), 1520-1531; (c) Lauterbach, M.; Wipff, G.; Mark, A.; van Gunsteren, W. F. *Gazetta Chim. Italiana* **127** (1997), 699-708.

CHAPTER 3

RECOGNITION OF NEUTRAL MOLECULES BY CALIXARENES IN SOLUTION AND IN GAS PHASE

A. POCHINI and A. ARDUINI
Dipartimento di Chimica Organica e Industriale, Università di Parma, Parco Area delle Scienze 17/a, I-43100 Parma, Italy.

Moving from biological chemistry studies, organic chemists were attracted by molecules having convergent concave surfaces as potential model of receptor sites *e.g.* of enzymes.[1] Using the cavity as binding site of a substrate (guest) this receptor (host) can perform a structural recognition in the complexation process on the base of the structural complementarity of the two molecular species. The strength of binding is controlled by the kinds and the numbers of simultaneous interactions, by the preorganisation of the system and by the solvation effects.

On this base Cram defines **cavitand** as "host compounds whose enforced concave surfaces are large enough to form complexes with organic guests without steric inhibition of either the complexation or the decomplexation process. Caviplexes are their complexes".[2] In general cavitand are open molecular systems with a vase-like structure in which strong competition between the guest and the solvent is observed. Consequently the complexation of organic guests in lipophylic solvents generally requires solvents which, because of their molecular size, cannot enter the cavity.[3]

An other class of receptor is characterised by a closed cavity, generally non-collapsible **molecular cages**. These hosts contains portals, through which guests can pass in either direction but with steric barriers. By this way steric inhibition of either the complexation or the de-complexation process occurs and the extreme form of complex stability is in carceplexes in which the guest can leaves the cavity only after the breaking of covalent bonds of the host (carcerand). In this way constrictive binding of the guest occurs. The closed cavity of the host can protect a reactive guest thus limiting also its degree of freedom.[3a, 4]

The different kinetic of the complexation process in the two class of receptors determines the field of application of these systems. Cavitands are useful in sensors preparation and in molecular separations whereas cages can be utilised in the fields of materials, molecular devices preparation and drug delivery.

3.1. Calix[4]arene Cavitands

One of the most attractive features of calix[4]arenes is their cuplike structure, which has been observed both in the solid state and in solution and from which this class of macrocycles derives the name "calixarenes".[5] In the cone conformation of calix[4]arenes a cavity is present which can host neutral guest molecules of complementary size. Several inclusion complexes of organic neutral molecules and p-alkylcalix[4]arenes have been isolated and the X-ray crystal structure determined. Starting from the first crystal structure of *endo*-cavity complex observed, *i. e.*. **p-*tert*-butylcalix[4]arene ⊂ toluene** complex, [6] systematic studies on the selective recognition of aromatic guests in the solid state were performed.[see Cap. 7] Moreover *endo*-calix complexes with p-*tert*-butylcalix[4]arene or its derivatives have been observed in the solid state not only with the aromatic molecules but also with other guests such as acetonitrile, methanol and methylene chloride.[7]

Moving from these very promising results, attempts to exploit the π-donor apolar cavity of calix[4]arenes for the complexation of organic neutral molecules in solution were performed. Positive results were initially obtained using water soluble hosts where hydrophobic effects drive the complexation.[5]

On the contrary, using apolar organic solvents the binding ability of *e.g.* p-*tert*-butylcalix[4]arene toward these guests are very poor. Cavitands derived form resorcinarene described by Cram in the eighties are nevertheless able to bind organic guests in apolar media. Although no systematic studies on the effect of structural parameters of those hosts on their complexation efficiency were performed, examination of their structure suggested the importance of rigidity to observe complexation. In particular, rigid receptors were obtained by introducing bridges or in specific cases also extending the apolar cavity of the cyclophane frame.[3a] Being very low the intermolecular interactions involved in the host-guest complex formation, the possibility to stabilise the complex increasing the preorganisation of the hosts both in apolar organic media and in the gas-phase was studied. The first attempt was the reduction of the mobility of the host blocking the *cone-to-cone* ring inversion of the calix.

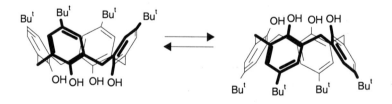

Figure 1. *Cone-to-cone* ring inversion of p-*tert*-butylcalix[4]arene.

The introduction of bulky substituent at the narrow (lower) rim of the calix prevents the interconversion among the four possible stereoisomers: *cone, partial cone, 1,2-alternate* and *1,3-alternate*.[5] So choosing the reaction conditions the tetraalkoxy derivatives of p-alkylcalix[4]arenes in *cone* conformation were prepared but no complexing ability toward neutral organic molcules was observed.

However, most of the tetra-O-alkylated calix[4]arene *cone* isomers adopt a *flattened cone* (pinched cone) [8] conformation in the solid state,[9] showing a C_{2v} symmetry. This conformation has also been shown to be the most stable by molecular modeling.[10] In solution, a residual conformational mobility of the *cone* isomer still exists, and the C_{4v} symmetry, usually observed in the ^{1}H NMR spectra of these compounds, is considered to be the result of a fast interchange between two C_{2v} structures. (see Figure 2).

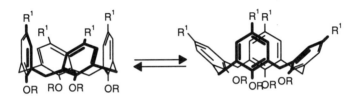

Figure 2. Dynamic stereochemistry of tetraalkoxycalix[4]arenes in the *cone* conformation.

The direct experimental evidence of this residual conformational flexibility was obtained [11] by studying the temperature dependence of ^{1}H NMR spectra of *cone* tetrakis(n-octyloxy)calix[4]arene (**1**, see Figure 6) which, in CD$_2$Cl$_2$, gives a spectrum compatible with a C_{4v} symmetry at room temperature. Decreasing the temperature to 213 K the spectrum of **1** changes, indicating a C_{2v} structure. These changes clearly show the freezing of the molecular motion into a preferred *flattened cone* conformation.

On the base of these results we hypothesise that the lack of complexing efficiency toward neutral organic molecules could be partially due to this residual mobility of these hosts.

CPK molecular models show that an attractive target for rigid hosts having additional binding sites on the *cone* structure is the linkage, at the upper rim of the calix, of the diametrical aromatic nuclei with rigid bridges containing π-donor groups.

Therefore a series of calix[4]arene cavitands bearing 2,4-*hexadiynyl* or p-*xylyl* moieties (*e. g.* **2** and **3**) was synthesised and the complexing properties toward neutral organic molecules studied and compared with those of non-bridged analogues. In order to enhance the host-guest interactions and prevent solvation of both host and guest, the complexing abilities were evaluated in the gas phase. The absolute and relative complexation efficiency was thus studied using Desorption Chemical Ionisation Mass Spectrometry.[12]

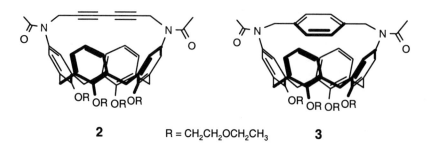

Figure 3. Calix[4]arene cavitands able to recognize guests in the gas phase.

Interestingly it emerged that the existence of a rigid cavity seems to be essential requisite to observe strong gas-phase supramolecular interactions, and the complexation efficiency strongly depends on the structure and length of the bridge present at the upper rim.

Table 1. Calculated gas-phase relative complexation constants for hosts **2** and **3** toward different classes of guests (referred to CH_3COOEt).

Guests	2	3
CH_3COOi-Pr	2.3	2.1
$CH_3COCH_2CH_3$	1.7	1.8
CH_3COOt-Bu	1.1	1.3
CH_3COOEt	1	1
CH_3COOn-Pr	0.29	0.48
CH_3CN	0.82	0.58
t-BuOH	0.76	0.28
i-PrOH	0.40	0.06
n-PrOH	0.08	0.03
C_6H_6	< 0.02	< 0.01

The cavitands exhibiting the highest complexation efficiency turn out to be also quite selective among the candidate guests (see Table 1). In fact **2** and **3** give extensive complexation (60-90%) with esters, ketones and acetonitrile, but interact weakly with benzene and alcohols. Thus the complex stability seems to be determined by multiple interactions between the acidic methyl hydrogen of the guest (α to the electron-withdrawing group) and the π-electrons of both cavity and bridge.

However, the binding ability experienced in the gas-phase by receptors **2** and **3** could not be demonstrated also in solution where solvation effects can occur. Therefore we designed and synthesised new cavitands (*e.g.* **4**) to improve the complexation ability by introducing more efficient additional binding sites, using pyridine containing bridges. By comparing the data obtained with **4** with those of the *iso*-phthaloyl analogue **5**, it was possible to investigate the role of the pyridine basic group located in close proximity to the calixarene cavity on complexation.[13]

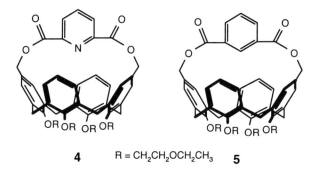

Figure 4. Upper rim bridged calix[4]arene derivatives in the cone conformation.

Nitromethane, malononitrile and acetonitrile were chosen as guests and the complexes studied by ^{1}H NMR spectroscopy in CDCl$_3$ and CCl$_4$.[13] Significant upfield shifts of the methyl or methylene protons of the guest were observed by adding the guest to a solution of host **4**. This indicates that the acidic protons of the guest interact with the π electrons of the host cavity. In fact, any interaction of these protons involving only the nitrogen of the pyridine moiety and the oxygen atoms present in the bridge should result in a downfield shift of these signals. By applying the continuous variation method a 1:1 stoichiometry for the complexes was established.

The *iso*-phthalic derivative **5** shows no complexation for any of the guests studied, clearly demonstrating that the presence of the pyridine nitrogen donor atom in the bridge is essential for the efficiency of the receptor.

From the values of the stability constants listed in Table 2 it emerges that the acidity of the guest, its shape and the polarity of the solvent affect the degree of complexation.

Table 2. Associations constants (K_{ass}, M^{-1}) for 1:1 complexation of bridged calix[4]arene with neutral guests at 300 K.

Calix[4]arene	CH_3CN CCl_4	CH_3NO_2 CCl_4	$CH_2(CN)_2$ $CDCl_3$
4	36±10	57±5	79±20
5	a	a	a

[a] no significant variation of the chemical shift observed.

As expected, the stability of the complexes is higher in the less polar CCl_4 and increases with the acidity of the guest (malononitrile > nitromethane > acetonitrile), while it is less sensitive (compared with the biscrown series, see ahead) to the steric demand of the guest.

The structures of the complexes $CH_2(CN)_2 \subset 4$ and $CH_3NO_2 \subset 4$ obtained from the crystallographic study are reported in Fig 5.[13]

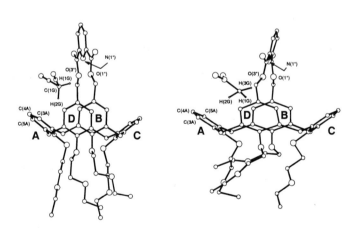

Figure 5. *X-ray* crystal structure of the $CH_2(CN)_2 \subset 4$ and $CH_3NO_2 \subset 4$ inclusion complexes.

The most interesting and common feature of the two complexes is that both guests are held by the host *via* the co-operation of two different hydrogen bonds: one "multifurcated" with the pyridine bridge and the other with the aromatic ring of the calixarene *via* a CH-π interaction. In particular, in the complex **CH$_2$(CN)$_2$ $\subset$ 4** the orientation of the pyridine bridging group with respect to the aromatic nuclei of the calixarene allows the host to act as "tongs" towards the malononitrile guest, which is complexed *via* a trifurcate hydrogen bond involving the hydrogen atom H(1G) of the malononitrile and the O(1*), N(1*), O(3*) acceptor atoms of the biscarbonyl pyridine bridge. The guest forms a weaker bifurcate hydrogen bond between H(2G) and the π orbitals on C(4A) and C(5A) atoms on the phenolic ring A with a bond distance of 2.85 and 2.87 Å respectively. The structural properties of **CH$_3$NO$_2$ $\subset$ 4** complex are significantly different. The calixarene host shows a more symmetric and less distorted *flattened cone* conformation, whereas the biscarbonyl pyridine group is more bent towards the reference plane of the calixarene. The complexation of the guest occurs mainly *via* a bifurcated hydrogen bond involving the H(3G) of the nitromethane and the two acceptor atoms O(3*) and N(1*) of the biscarbonyl pyridine bridge. The guest is thus blocked in the observed position by the additional hydrogen bond between H(2G) and the π cloud of aromatic nucleus A (distance H(2G) - C(3A) = 2.71 Å). The X-ray crystal structure then confirms the participation in the binding process of both "hard" donor groups on the bridge and the "soft" aromatic cavity.

As evidenced by the negative results obtained with the *iso*-phtalic derivative **5** the efficiency of the host **4** is due to the presence of the pyridino additional binding site. With the aim to exploit the aromatic cavity of calix[4]arenes as the only binding sites for neutral organic molecules, and in order to reduce the conformational mobility of the *cone* conformer of these hosts other approaches were tried.

CPK molecular models show that an attractive target for a very rigid and not distorted calix[4]arene *cone* derivative is the linkage of two proximal phenol rings with short diethylene glycol bridges at the lower rim.

It was therefore initially developed an indirect multi-step procedure for the selective 1,2 (proximal) functionalization of calix[4]arenes at the lower rim, [14] but more recently we have found a direct one-step synthetic procedure to obtain a series of calix[4]arene biscrowns (**6-10**) where the conformational flexibility of the calix[4]arene cavitands (see Figure 6) can be tuned by varying the length of the bridging unit. [15]

In fact, while p-*tert*-butylcalix[4]arene-biscrown-5 (**10**) is flexible and adopts a *flattened cone* conformation in the solid state, [16] calix[4]arene-biscrown-3 derivatives (*e.g.* **8**) possess a rigid *cone* structure. The recognition properties of rigidified p-*tert*-butylcalix[4]arene-biscrown-3 (**7**) in solution were evaluated in apolar organic solvents (chloroform, carbon tetrachloride) using nitromethane as guest. [17] By adding variable amounts of guest to a solution of **7** a significant upfield shift of the CH protons of the guest and fast exchange conditions are observed. These shifts clearly show an interaction of the acidic protons of the guest with the π electrons of the calixarene cavity, while excluding a possible interaction of these protons with the crown-ether region which should result in a downfield shift of these signals.

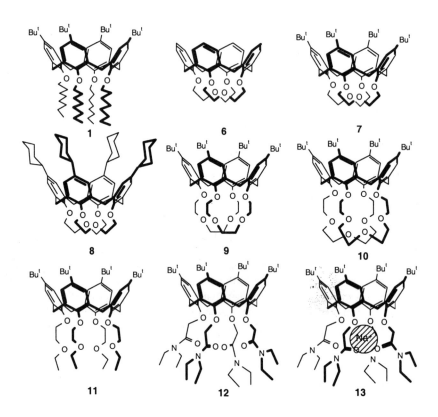

Figure 6. Calix[4]arene hosts having different rigidity.

The analysis of the complexation data shows the formation of 1:1 complex and provides quantitative information on the Host-Guest interactions (see Table 3). In order to gain further insight into the role of rigidity on the molecular recognition properties of calix[4]arene receptors we studied the behaviour of the more mobile tetrakis-(2-ethoxyethoxy)-p-*tert*-butylcalix[4]arene (**11**) and p-*tert*-butylcalix[4]arene-biscrown-5 (**10**). Interestingly, with both hosts no variation of the chemical shift of the guest is observed, while with **9** complexation does occur only in CCl$_4$, but with a lower binding constant than with the more rigid **7** (see Table 3).

Table 3. Associations constants (K_{ass}, M^{-1}) for 1:1 complexation of nitromethane and malononitrile by rigidified calix[4]arene at 300 K.

Calix[4]arene	CH_3NO_2		$CH_2(CN)_2$
	$CDCl_3$	CCl_4	$CDCl_3$
6	5±2	28±7	17±2
7	27±4	230±60	6±2
8	36±8	123±25	23±5
9	b	50±10	B
10	b	a	B
11	a	a	B
12	a	b	B
13	34±7	b	B

[a] no significant variation of the chemical shift observed; [b] not determined.

These results demonstrate the importance of rigidity in determining the complexation properties of the π donor cavity of calix[4]arenes. As expected, the groups present at the upper rim of calix[4]arene-biscrown-3 strongly affect the K_{ass} values. In fact, increasing the extension of the cavity the complexation efficiency increases, as deduced by comparing the K_{ass} values of **6**, **7**, and **8**. In addition, malononitrile which is more acidic than nitromethane, is bound at a lesser extent, indicating that not only the acidity but also the shape of the guest included plays an important role and that steric hindrance strongly reduces the binding.

The X-ray crystal structure of the **$CH_3NO_2 \subset 8$** complex (Figure 7) [17] further supports the data obtained in solution and proves the symmetrical *cone* structure of the host.

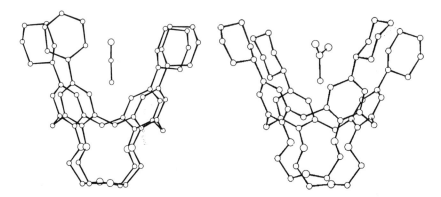

Figure 7. *X-ray* crystal structure of the **$CH_3CN \subset 8$** and **$CH_3NO_2 \subset 8$** complexes.

In the solid state the guest molecule lies inside the cavity and orients its N-C bond along the axis of the *cone* whereas the CH$_3$ group faces the aromatic nuclei of the host.

Another interesting observation, which confirms the importance of rigidity and, consequently, of the preorganization of the *cone* conformer of calix[4]arenes in determining their complexation ability, is obtained by the comparison between *p-tert*-butylcalix[4]arene tetraamide (**12**) and its sodium picrate complex (**13**) [18] in the complexation of nitromethane in CDCl$_3$. In fact, while the conformationally mobile **12** does not show any significant complexation, its rigid sodium complex **13**, through an allosteric effect, strongly binds CH$_3$NO$_2$ (see Table 3). Through these results the possibility of rigidifying these hosts not only using covalent bonds but also with other linkages was shown.

Further confirm came from research of Stibor and his co-workers, which showed the possibility of utilising as efficient hosts, for guests bearing acid CH, calix[4]arenes partially alkylated at the lower rim, *e.g.* 1,3-dialkoxy derivatives **17** (see Table 4). [19]

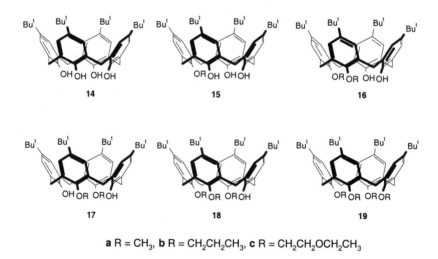

a R = CH$_3$, **b** R = CH$_2$CH$_2$CH$_3$, **c** R = CH$_2$CH$_2$OCH$_2$CH$_3$

Figure 8. *Alkoxy derivatives of p-tert-butylcalix[4]arene.*

The rationalisation of these results can be found in the studies on the conformational distribution and interconversion of lower rim partially alkylated calix[4]arenes performed by Reinhoudt and his group.[20] In particular the conformation of the partially methylated derivatives (**15a-18a**) are far less mobile than either the tetrahydroxy (**14**) or the tetramethoxycalix[4]arene (**19a**) *e. g.* the 1,3-

dimethoxy ether (**17a**) adopts a *cone* conformation both in solution and in the solid state and does not show any sign of coalescence in the NMR spectrum at temperatures up to 125 °C. The order of barriers calculated for the *cone-inverted cone* interconversion is monomethoxy (**15a**) > 1,2-dimethoxy (**16a**) > 1,3-dimethoxy (**17a**) > trimethoxy derivative (**18a**).

Further experimental data in the evaluation of the energy barriers comes from Böhmer's studies on the racemization of partially methylated chiral calix[4]arenes (see Figure 9). Calix[4]arenes (*e.g.* **20**) bearing on the macrocycle *meta*-substituted phenols such as 3,4-dimethylphenol are chiral in their *cone* conformation and their *cone-to-cone* inversion converts one enantiomer into its mirror image.[21] Thus investigations of this thermal racemization by measurement of rate constants as a function of temperature should enable the experimental determination of the energy barriers for the chiral derivatives **20, 21** and **22**.

An energy barrier of $\Delta G^{\ddagger}$ = 13.4 kcal mol-1 at 291 K was found for **20** in CDCl$_3$; this value is slightly lower than that reported for p-methylcalix[4]arene ($\Delta G^{\ddagger}$ = 14.6 kcal mol^{-1} at 323 K) under analogous conditions. As expected, higher values for $\Delta G^{\ddagger}$ are found for **21** and **22** and in agreement with the theoretical expectations the barrier is higher for the monomethyl ether **21**, with three intramolecular hydrogen bonds, ($\Delta G^{\ddagger}$ = 24.3 kcal mol^{-1}) than for the dimethyl ether **22**, with two intramolecular hydrogen bonds, ($\Delta G^{\ddagger}$ = 23.3 kcal mol^{-1}). This difference is more pronounced in $\Delta H^{\ddagger}$ than in $\Delta G^{\ddagger}$ due to a strong compensation from $\Delta S^{\ddagger}$.[21]

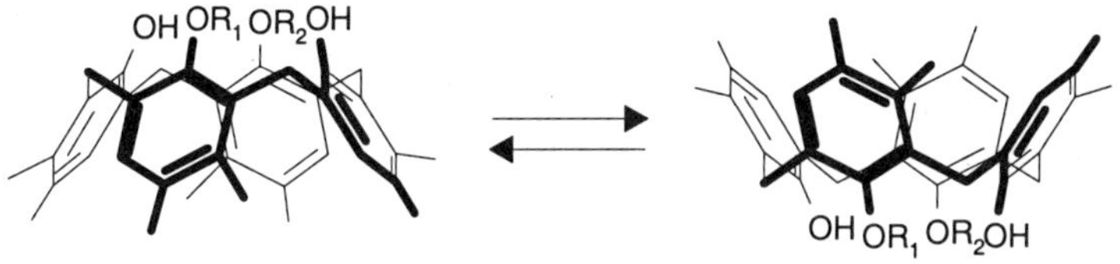

20, $R_1 = R_2 = H$
21, $R_1 = CH_3$ $R_2 = H$
22, $R_1 = R_2 = CH_3$

Figure 9. *Cone-to-cone* ring inversion of dimethylcalix[4]arene derivatives.

So with these hosts the rigidity derives from a combination of hydrogen bonds and steric repulsions. Using carbon tetrachloride as solvent and *e. g.* acetonitrile as guest no complexation was observed with tetraalkoxy derivative of p-*tert*-butylcalix[4]arene **19**. An association constant at 298 °K of 39 M^{-1} with tetrahydroxy derivative **14**, which increase up to 157 M^{-1} with the 1,3-

diethoxyethoxy **17c** and to 232 M^{-1} with the monoethoxyethoxy **15c** derivative, was observed (see Table 4).[19] These results show that the most efficient host is the most rigid, confirming the importance of the entropic term in determining the association constant in the host-guest interaction between an aromatic cavity and a neutral organic molecule.

Table 4. Association constants (K_{ass}, M^{-1}) of the 1:1 complexes of calix[4]arenes **14-17** with neutral guests in CCl$_4$ at 298 K.

Calix[4]arenes	CH$_3$CN	ClCH$_2$CN	C$_2$H$_5$CN	CH$_3$NO$_2$	C$_2$H$_5$NO$_2$
14	39	27	18	52	21
15c	232	109	61	7.5	284
17c	157	-	19	129	19
17a	80	65	14	76	-
17b	152	94	20	132	20

Probably also the results obtained by Fukazawa *et al.*, who studied the binding ability of monomethoxy-monodeoxy-p-*tert*butylcalix[4]arene derivative towards guests having acidic CH groups, can be interpreted in this context. A general trend observed with this rigid host is that, in CCl$_4$ solution, their binding efficiency parallels the acidity of the CH group of the organic guest.[22]

In all these studies no evidence of complexation of aromatic molecules, using apolar solvents, were obtained further supporting the importance of the presence of acid CH in the guest. However, preliminary X-ray solid-state studies with guests having different acidity (*e.g.* **CH$_3$CN $\subset$ 8** and **CH$_3$NO$_2$ $\subset$ 8**) have shown that the these CH-π interactions are directional although the distance between the carbon of the guest and the centre of the aromatic nuclei of the host are not affected by the acidity of the guests. (see Fig. 7). [23] These data point out the need of further studies to understand these specific interactions.

A general goal of host-guest chemistry is also to transfer molecular recognition phenomena experienced by synthetic hosts in order to devise systems dealing with a specific function, especially with a view towards practical and technical applications.

Recently the application of Quartz Crystal Microbalance (QCM) to quantitatively determine organic analytes have received considerable attention.[24]

The QCM technique is based on the ability of a layer, coating the oscillating quartz crystal, to absorb in a fast and reversible manner, gaseous analytes. The interaction between the coating and the analyte results in a mass change of the coated quartz, which produces a frequency drop. The frequency decreases and the mass changes are quantitatively related by the Sauerbray equation ($-\Delta f = k \Delta m$).

The choice of the active coating was however made amongst the absorbents used in gas chromatography which, operating through non specific absorbent-analyte interactions, experience low selectivity toward organic analytes.[25] In order to gain selectivity, the active coatings have recently been chosen among simple molecular receptors widely used in supramolecular chemistry (including calixarenes).[26] Nevertheless, whether these coatings sense vapours of organic analytes through molecular recognition processes is still open to discussion[27].

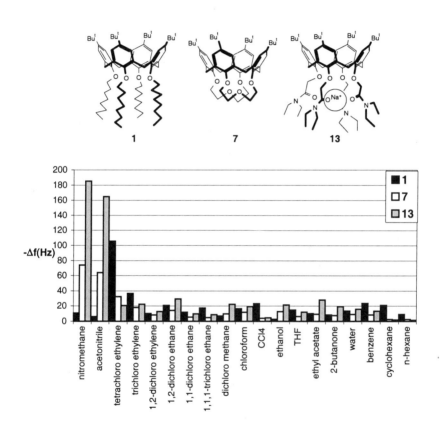

Figure 10. Sensing ability of calix[4]arene having different rigidity as coatings of QCM trasducers toward different classes of analytes.

Within this context, the sensing ability of a series of functionalized calixarene-based coatings towards different classes of organic analytes was tested, using QCM.[28]

The data reported in Figure 10, though preliminary, show some interesting features. In fact, when the coating is made with flexible calix[4]arenes (**1**), which probably sense analytes through non specific interactions, a similar behaviour to that observed with aromatic polymers coatings was verified. On the contrary, using more rigid calixarenes (*e. g.* **7** or **13**) a significant -Δf was observed for analytes having acidic CH groups such as nitromethane and acetonitrile and a low Δf for water and ethanol.

These data, which parallel those obtained in apolar media, strongly suggest that these coatings are able to sense analytes through molecular recognition processes.

3.2. Hybrid Calix[4]arenes

Hybrid macrocyclic receptors, characterised by the presence of binding units of different nature, *e. g.* polar side-arms and hydrophobic cavities or surfaces, are useful systems for recognition processes. Particularly interesting are the calix[4]arene receptors containing amino acid residues.

By the introduction of four or two (in diametrical position) L-alanine methyl ester or L-alanyl-L-alanine methyl ester residues at the upper rim of tetrapropoxycalix[4]arene in the *cone* conformation new chiral hosts (*e. g.* **23a** and **23b**) were obtained.[29] Reaction of these compounds with hydrazine gave the corresponding hydrazido derivatives (**23c**, **23d**). Preliminary binding studies indicates that the difunctionalised hydrazido compound **23c** and **23d** are able, in organic media, to extract D-alanyl-D-alanine.

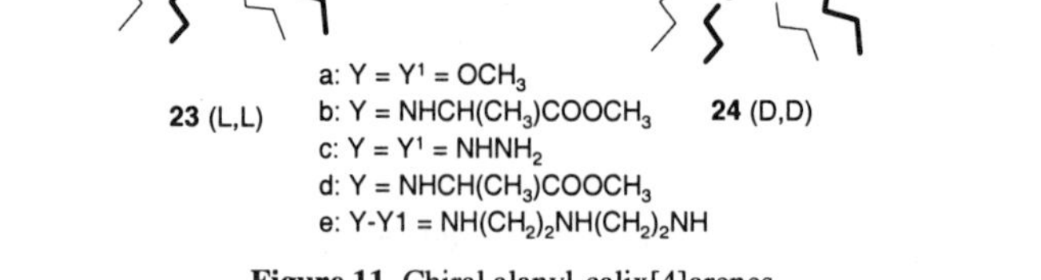

23 (L,L)
a: Y = Y¹ = OCH₃
b: Y = NHCH(CH₃)COOCH₃
c: Y = Y¹ = NHNH₂
d: Y = NHCH(CH₃)COOCH₃
e: Y-Y1 = NH(CH₂)₂NH(CH₂)₂NH

24 (D,D)

Figure 11. Chiral alanyl-calix[4]arenes.

Linking two amino acid residues with a diethylentriamino bridge the new hosts **23e** and **24e** were synthesised.[30] Using solid N-acetyl-D-alanyl-D-alanine as guest and a CDCl₃ solution of **23e**, a soluble 1 : 1 complex forms with an association constant of $>10^5$ M^{-1}. Using soluble N-lauroyl-D-alanyl-D-alanine homogeneous titration with host **23e** confirmed these data, whereas N-lauroyl-D-alanine gave an association constant of 11000 M^{-1}.

On the base of these data a comparison of the antibacterial activity of this receptor with that of vancomycin antibiotic was performed. The action of vancomycin proceeds in fact by binding to the cell wall of Gram-positive bacteria, terminating in the sequence D-alanyl-D-alanine, thus inhibiting the growing of the cell wall and causing the cell lysis.

The binding of D-ala-D-ala was indirectly verified in this preliminary antibacterial activity evaluation. Receptor **23e** shows a good anti Gram-positive activity although slightly inferior to vancomycin.[30]

3.3. Larger Calixarenes

The larger members of this class of macrocycles (calix[5], -[6] and –[8]arene) have long been neglected. The reason for this is probably due to their higher conformational mobility which render difficult the problem of fixing these macrocycles in a given conformation or orienting functional groups and binding sites. However, recently the need of receptors for larger species has stimulated the research of methods for their selective functionalisation on both rims.

In fact, a tremendous increase of interest in large calixarenes rose after the discovery of Atwood and Shinkai who discovered the possibility separating C_{60} from carbon soot exploiting the molecular recognition properties of p-*tert*-butylcalix[8]arene (**25**) and thus devising a convenient method to separate C_{60} from C_{70}.[31].

The reason why C_{60} is included in preference to C_{70} and why the complex precipitate so efficiently from a toluene solution was rationalised by Shinkai *et al.* in terms of entropy effects.[32] In fact, while C_{60}, which is ball shaped, can maintain its rotational freedom even in the solid state, thus justifying a rotational freedom also in the complex with p-*tert*-butylcalix[8]arene (**25**), the formation of the complex between **25** and C_{70}, which is rugby-ball shaped, should result in a loss of rotational freedom. It is therefore reasonable to assume that the rotational difference is reflected by the entropy term so that the K_{ass} for the C_{60} complex formation can become larger than that for C_{70}.

Most of the studies in this field, have thus been devoted to establishing the structure, the nature of the interactions involved in calixarene-fullerene conjugates, [33] and the properties of the novel supramolecular architectures and nanostructures obtained both in solution an in the solid state.

Therefore several calix[n]arenes including homooxacalixarenes [34] and other cyclophanes [35] were screened as potential hosts for the inclusion of fullerenes (C_{60} and C_{70}) or other icosahedral clusters having comparable dimension (diameter) [36]. The solid state studies carried out in this topic have been exhaustively discussed in the chapter 7 by Ugozzoli; in the current chapter only the studies carried out in solution are reviewed.

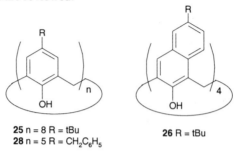

25 n = 8 R = tBu
28 n = 5 R = CH$_2$C$_6$H$_5$

26 R = tBu

Figure 12. Calixarenes for the C$_{60}$ fullerene complexation.

With the exception of a calix[4]naphtalene (**26**), which possesses an extended cavity and is able to bind C$_{60}$ with a K$_{ass}$ = 6920 dm^{-3} mol^{-1} in CS$_2$, the size of calix[4]arenes is considered to be too small to interact efficiently and form *endo*-cavity inclusion complexes with fullerenes both in solution and in the solid state.[37]

In an attempt to rationalise the results obtained on the complexation properties of several calix[n]arenes toward fullerenes in organic solvents, Shinkai [38] pointed out that a primary prerequisite for C$_{60}$ inclusion is that the OH groups at the lower rim of the macrocycle are not substituted, that is, sufficiently preorganised through intramolecular hydrogen bonding. Moreover, as additional prerequisite, considering that the major driving forces involved in the formation of these host-guest complexes are π–π (including charge-transfer) and/or solvophobic type, the macrocycle should be large enough to allow its aromatic nuclei to assume an inclination suitable for the maximisation of multi-point contact formation with the globular shaped, weak electron-acceptor C$_{60}$.

The groups present at the upper rim of the calixarene skeleton also play an important role as elegantly showed by Fukazawa *et al.* [39] In fact, studying the binding ability of upper rim functionalised calix[5]arenes (**27a-c**) toward C$_{60}$, it was found that compound **27a**, having two iodine atoms as additional binding sites, has the highest association constant for the a 1:1 complex with C$_{60}$. As shown in Table 5 the association constant values, determined by absorption spectroscopy, are also strongly solvent dependent, and this was mainly ascribed to the different solubility of the guest and its different solvation in the different solvents studied. It was then concluded that the more weakly solvated guest is more strongly bound in the host.

A further insight into the role played by solvation phenomena on the *endo*-cavity complex formation was achieved by Isaacs *et al.* who studied the partial molar volume change associated with complexation of C$_{60}$ by p-benzylcalix[5]arene

(28) (see Fig. 12) in toluene (+195 cm^3 mol^{-1}) using high-precision densitometry techniques and concluded that the host cavity is solvated by two toluene molecules that are displaced by the guest upon complex formation [40].

Table 5. Association constants (K_{ass}, M^{-1}) for the 1:1 complexes of calix[5]arenes (**27a-c**) with C$_{60}$.

Calix[5]arenes	Toluene	Benzene	CS$_2$	Dichlorobenzene
27a	2120	1840	660	308
27b	1673	1507	600	277
27c	588	459	284	207

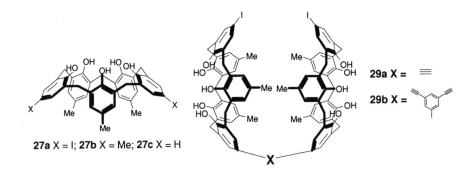

Figure 13. Calix[5]arene receptors for fullerenes.

The UV-Vis spectra of toluene solutions of C$_{60}$ with increasing ratios of **28** indicates that together with a 1:1 Host-Guest complex that forms with a K_{ass} = 2800 ± 200 dm^{-3} mol^{-1} for low concentration of p-benzylcalix[5]arene, a second complex having an Host-Guest stoichiometry of 2:1 forms, with a K_{ass} = 230 dm^{-3} mol^{-1} at higher concentration of host.[41]

The 2:1 host-guest solid state structure of **27a** with C$_{60}$ was the starting point to design new receptors with a well defined cavity shape and how two calix[5]arenes should be linked, covalently [42] or through self-assembly processes [43], to give more efficient hosts for the selective recognition of fullerene

Interestingly the double calix[5]arenes (**29a, b**) binds C$_{70}$ preferentially to C$_{60}$ and while the highest C$_{70}$/C$_{60}$ selectivity (10.2) was observed with **29a** in toluene, the larger association constant (K_{ass} = 163 ± 16 x 10^3) was observed using **29b** as host in toluene. Through ^{13}C NMR experiments and molecular-mechanics calculations studies on the complex between **29b** and C$_{70}$ it was also possible to conclude that the poles of the bound guest reside the deepest within the host cavity.

3.4. Molecular Cages.

Following the Cram's pioneering work on the synthesis of carcerands and emicarcerands bridging the upper rim of two resorcinarenes, [3a] several upper rim-upper rim covalent linked double calix[4]arenes were prepared but no evidence for neutral organic molecules complex formation was obtained.

More productive for the synthesis of molecular cages, able to recognise neutral molecules, was "the modular approach" pursued by Reinhoudt and his group which combines calix[4]arenes with different molecular building blocks and in particular with resorc[4]arenes.[44] The scheme of the synthesis of the carcerand (**31**) is reported in Figure 14 [45]

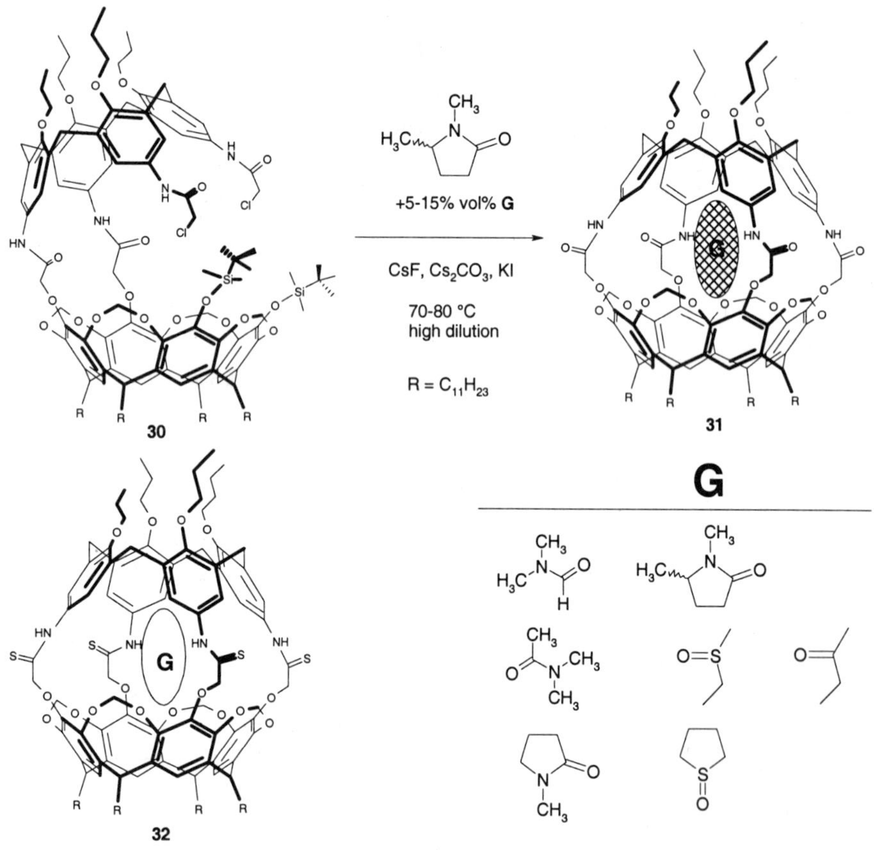

Figure 14. *Calix[4]arene-resorc[4]arene carceplexes* and preferred orientation of the guest in the cavity.

The key step to prepare the carceplexes is the closure of the final two bridges of **30**, which must be performed in solvents such as the amides, e. g. N,N-dimethylacetamide and N-methylpyrrolidin-2-one, and sulphoxides, e. g. ethyl methyl sulphoxide, giving solvent included carceplexes. The limit of the method is necessity for use highly polar solvents in this step. To include less apolar molecule another method, called doped inclusion, was developed. In this case 1,5-dimethyl-2-pyrrolidinone, which itself is a poor guest for the incarceration, is used as solvent and potential guests are added in 5-15 vol %.In several doped inclusion competition experiments the templating ability of different guests was studied and in Table 6 the data obtained using different guests during the synthesis of calix[4]arene-resorc[4]arene carceplexes by doped inclusion are reported.

From Table 6 it is clear that DMA is the best template for the synthesis of this carcerand. These results are in agreement with the studies performed by Cram and Sherman, which demonstrated the essential prerequisite of a guest template reaction for success in carcerands synthesis Since empty carcerand are not observed and this molecular cage can only be formed when the guest occupies the cavity, the templating ability is an indirect measure of the association strength between the host and the guest.[46]

The best interaction between this carcerand and DMA might be due to the guest polarity as well as the size and shape of the guest.

Table 6. Templating ability of potential guests during the synthesis of **31**-based carceplexes by doped inclusion.

Guests	Templating ability [a]	Yield (%) [b]
DMA	100	27
DMSO	63	16 [c]
DMF	27	13 [c]
2-Butanone	27	16

[a] DMA is set at 100. [b] Isolated carceplex when only one guest is used during doped inclusion. [c] Yield of deuterated guest.

In order to extend the number of different calix[4]arene-based carcerands, the amide bridges of the different carceplexes were converted in thioamide bridges in quantitative yield by means of Lawesson's reagent in refluxing xylene obtaining carceplexes **32**.

An interesting and exclusive feature of these carcerands is the nonsymmetric cavity structure, which results in varying orientation of the incarcerated guest ("carceroisomerism"). The dynamic properties of the incarcerated guests were examined by 2D NMR spectroscopy. Whereas for some guests a preference for one orientation inside the carcerand was observed, while DMA, NMP, and ethyl methyl sulfoxide inside carcerands two different orientations were present. The experimentally determined preferred orientation of guests inside calix[4]arene-based carcerand is reported in Figure 13. The energy barrier for the various orientations were determined and it was verified that the energy barriers are higher for the thiacarcerands **32** than for the corresponding carcerand with amide bridges **31**. The larger barrier for the thiacarceplexes **32** might indicate that the cavity inside the calix[4]arene-based thiacarceplexes is smaller than the corresponding amide-bridged carceplexes. Molecular modelling studies seem to support this hypothesis.

These supramolecular systems can be considered as a new approach towards a molecular switch, useful for the preparation of "storage devices".

The preparation of molecular cages by the linkage of the upper rim of two calix subunits can be performed not only using covalent bonds but also by self assembly. The preparation, the structure and the properties of these molecular capsules will be discussed in detail in Chapter 9. Here only some complexing properties of these systems will be discussed.

Upper rim arylureas derivatives of tetraalkoxy calix[4]arenes **33** dimerize reversibly in apolar organic solvents and through the formation of sixteen hydrogen bonds create a molecular capsule (**33·33**). [47] This dimerization is driven by the decrease of enthalpy due to the formation of intermolecular hydrogen bonds between the urea functions in a cyclic "head-to-tail" arrangement and by a slight loss of entropy (see Figure 15).

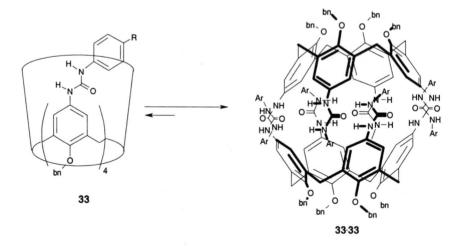

Figure 15. Arylurea derivatives of tetraalkoxycalix[4]arene in the cone conformations

Small molecule guests of suitable size and shape (*e. g.* benzene) are reversibly encapsulated in the cavity of these hosts on a time scale that is low with regard to

NMR spectroscopic measurements. Using a p-fluoro aryl urea derivative (**33a**, R = F) as host and [D_{10}] p-xylene as solvent binding of several guests were studied by ^{1}H NMR spectroscopy. When excess benzene is added to the solution of **33a** a new singlet appears at δ = 4.02 in the NMR spectrum and grows over the course of about 40 minutes. Integration indicates an approximately 1:1 ratio of benzene to the capsule **33a·33a** and the half-life for the uptake is about 8 minutes. The association constant in this solvent of benzene with **33a·33a** is 230 M^{-1}.[48]

^{1}H NMR experiments gave useful information on the orientation of asymmetric guests in the cavity of the molecular capsule. For example the *ortho-* and *meta-*hydrogen atoms of encapsulated fluorobenzene are shifted more, about 2 ppm, than are the *para*-hydrogen atoms. This is the evidence of a specific orientation of fluorobenzene within the cavity. The *ortho-* and *meta*-hydrogen atoms are directed towards the π -electron systems of the phenyl ring of the calix[4]arene, while the *para*-hydrogen atom is directed towards the encircling urea groups.

Competition studies with several guests were also undertaken and the affinity of the calix[4]arene dimer **33a·33a** for these (relative to benzene) is reported in Table 7.

Recently in a general study Mecozzi and Rebek demonstrated that molecular recognition through encapsulation processes is largely determined by the volumes of the guest and the host cavity. In particular the binding of molecules of suitable dimensions in the internal cavity of a molecular receptor in solution can be expected when the packing coefficient (PC), which is the ratio of the guest volume to the host cavity volume, is within the range of 0.55 ± 0.09.[49] Other different packing coefficients can be reached if the complex is stabilised by strong intermolecular forces such as hydrogen bonds.

Table 7. Relative affinities (25°C) of **33a·33a** for guests in competition with benzene.

Guests	
Benzene	1.0
Fluorobenzene	2.6
p-Difluorobenzene	5.8
Chlorobenzene	0.30
Toluene	<0.1
Phenol	0.83
Aniline	0.32
Pyrazine	3.2
Pyridine	1.2

The volume calculated for the cavity of the dimers **33·33** is 190 Å^3 and using the PC = 0.55 rule cubane was identified as excellent guest (V = 103 Å, PC = 0.54). The calixarene capsule **33·33** also binds smaller guests such as 1,4-difluorobenzene (PC

= 0.44) and pyrazine (PC = 0.38).The reason for the binding of these smaller guests can be found in specific interactions of these guests with the cavity. The partially positive CH bonds of both pyrazine and 1,4-difluorobenzene are directed toward the π– surfaces at the poles of the capsule and the basic heteroatoms are directed toward the acid urea hydrogens at the equator of the cavity. These specific interactions determine whether or not the guest will be encapsulated and how far from the PC = 0.55 solution it can be accomodated.

As useful comparison the volume of the cavity of the calix[4]arene-based carcerand **31** previously discussed is 159 Å^3 and the PC = 0.55 rule can then explain the templating effect of different solvents. For example, the reaction in N-methyl-2-pyrrolidinone (PC = 0.58) gives a yield of 50%, but when 1,5-dimethyl-2-pyrrolidinone (PC = 0.74) is used, the yield is reduced to <5%.

An alternative approach to capsule formation is the exploitation and matching of the geometrical coordination preferences of transition metals and the adaptability of calixarene frame. Shinkai and co-workers have in fact shown that two *cone* homooxacalix[3]arenes having 4-pyridyl groups at the upper rim do posses the appropriate geometry to be edge-to-edge dimerised *via* Pd(II) co-ordination. The dimeric capsule includes two dichloromethane molecules as guest.[50] In addition, the inner volume and symmetry of this cage match quite well that of fullerene C_{60} that is in fact included to form a 1:1 complex in $C_2D_2Cl_4$ solution. [51]

3.5. Conclusions

Recognition of neutral organic guests is based on the structural complementarity between the synthesised host and the guest. Surely the new tools, such as new prefabricated molecular sub-units, new functionalisations and the new methodologies for their assembly, will help in the design of new calixarene-based hosts for the recognition of specific guests.

However, further efforts are needed to understand how the intermolecular forces, responsible for the host guest interaction, operate in the recognition process. A typical example is the CH-aromatic interaction, which have been studied using more acidic *sp* and *sp*^{2} CH groups, whereas very few is known on the *sp*^{3} CH-aromatic interaction especially when these operate in a cooperative way with the several aromatic nuclei present in the cavity of a calixarene.

New efficient and selective cavitands, able to recognise neutral organic species have been synthesised and it is likely to foresee that these receptors will have an important role in the development of sensor technology. In addition, the preparation of new molecular cages and their complexes with neutral molecules will disclose new possibilities as molecular reactor or in the field of materials, information storage at molecular level.

3.6. References

1. Lehn, J.-M., *Supramolecular Chemistry: Concepts and Perspectives* (VCH, Weinheim, 1995).
2. Cram, D.J. *Nature,* **356** (1992), 29-36.
3. (a) Cram, D.J.; Cram, J.M., *Container Molecules and Their Guests* Monographs in Supramolecular Chemistry, ed. Stoddart, J.F., (The Royal Society of Chemistry, Cambridge 1994). (b) Collet, A. in *Comprehensive Supramolecular Chemistry*, ed. Vögtle, F. (Pergamon, Oxford, 1996) Vol.2, 325-365.
4. (a) Sherman, J.C.; Jasat, A., *Chem. Rev.* **99** (1999), 931-967. (b) Rebek, J. Jr., *Acc. Chem. Res.* **32** (1999), 278-286.
5. (a) Gutsche, C.D., *Calixarenes Revisited* Monographs in Supramolecular Chemistry, ed. Stoddart, J.F. (The Royal Society of Chemistry, Cambridge, 1998). (b) Böhmer, V., *Angew. Chem. Int. Ed. Engl.* **34** (1995), 713-745. (c) Pochini, A.; Ungaro, R. in *Comprehensive Supramolecular Chemistry*, ed. Vögtle, F. (Pergamon, Oxford, 1996) Vol.2, 103-142.
6. Andreetti, G.D.; Ungaro, R.; Pochini, A., *J. Chem. Soc., Chem. Commun.* (1979), 1005-1007.
7. See Ref. 5a, pag. 169.
8. For indirect evidence in solution see: Yamada, A.; Murase, T.; Kikukawa, K.; Arimura, T.; Shinkai, S., *J. Chem. Soc. Perkin Trans. 2,* (1991), 793-797.
9. Andreetti, G.D.; Ugozzoli, F.; Pochini, A.; Ungaro, R. in *Inclusion Compounds*, eds. Atwood, J.L.; Davies, J.E.; McNicol, D.D., (Oxford University Press, Oxford, 1991), Vol. 4.
10. Grootenhuis, P.D.J.; Kollman, P.A.; Groenen, L.C.; Reinhoudt, D.N.; van Hummel, G.J.; Ugozzoli, F.; Andreetti, G.D., *J. Am. Chem. Soc.* **112** (1990), 4165-4176.
11. Arduini, A.; Fabbi, M.; Mantovani, M.; Mirone, L.; Pochini, A.; Secchi, A.; Ungaro, R., *J. Org. Chem.* **60** (1995), 1454-1457.
12. Arduini, A.; Cantoni, M.; Graviani, E.; Pochini, A.; Secchi, A.; Sicuri, A.R.; Ungaro, R.; Vincenti, M., *Tetrahedron* **51** (1995), 599-606.
13. Arduini, A.; McGregor, W.M.; Pochini, A.; Secchi, A.; Ugozzoli, F.; Ungaro, R., *J. Org. Chem.* **61** (1996), 6881-6887.
14. Arduini, A.; Casnati, A.; Dodi, L.; Pochini, A.; Ungaro, R., *J. Chem. Soc., Chem. Commun.* (1990), 1597-1598.
15. Arduini, A.; Domiano, L.; Pochini, A.; Secchi, A.; Ungaro, R.; Ugozzoli, F.; Struck, O.; Verboom, W.; Reinhoudt, D.N., *Tetrahedron* **53** (1997), 3767-3776.
16. Pèpe, G.; Astier, J.-P.; Estienne, J.; Bressot, C.; Asfari, Z.; Vicens, J., *Acta Crystallogr., Sect. C* **51** (1995), 726-729.

17. Arduini, A.; McGregor, W.M.; Paganuzzi, D.; Pochini, A.; Ugozzoli, F.; Ungaro, R., *J. Chem. Soc:, Perkin Trans.* 2, (1996), 839-846.
18. Arduini, A.; Ghidini, E.; Pochini, A.; Ungaro, R.; Andreetti, G.D.; Calestani, L.; Ugozzoli, F., *J. Incl. Phenom. Mol. Recogn. Chem.* **6** (1988), 119-134.
19. Smirnov, S.; Sidorov, V.; Pinkhassik, E.; Havlicek, J.; Stibor, I., *Supramol. Chem.* **8** (1997), 187-196.
20. Van Hoorn, W.P.; Morshuis, M.G.H.; van Veggel, F.C.J.M.; Reinhoudt, D.N., *J. Phys. Chem. A* **102** (1998), 1130-1138.
21. Kusano, T.; Tabatabai M.; Okamoto, Y.; Böhmer, V., *J. Am. Chem. Soc.* **121** (1999), 3789-3790.
22. Yoshimura, K.; Fukazawa, Y., *Tetrahedron Lett.* **37** (1996), 1435-1438.
23. Arduini,A.; Pochini, A.; Ugozzoli, F. unpublished results.
24. (a) Grate, J.W.; Martin, S.J.; White, R.M., *Anal.Chem.* **65** (1993), 940A-948A. (b) Grate, J.W.; Martin, S.J.; White, R.M., *Anal. Chem.* **65** (1993), 987A-996A.
25. (a) Patrash, S.J.; Zellers, E.T, *Anal. Chem.* **65** (1993), 2055-2066. (b) McGill, A.R.; Abraham, M.H.; Grate, J.W., *Chemtech* **24** (1994), 27-37.
26. (a) Reinhoudt, D.N.; Gopel, W., *Science* **265** (1994), 1413. (b) Dickert, F.L.; Stathopulos, Rief, M., *Adv. Mat.* **8** (1996), 525. (c) Cygan, M.T.; Collins, G.E.; Dunbar, T.D.; Allara, D.L.; Gibbs, C.G.; Gutsche, C.D., *Anal. Chem.* **71** (1999), 142-148.
27. (a) Grate, J.W.; Patrash, S.J.; Abraham, M.H.; My Du, C., *Anal. Chem.* **68**, (1996), 913. (b) Hartmann,J.; Auge, J.; Lucklum, R.; Roster, S.; Hauptmann, P.; Adler, B.; Dalcanale, E., *Sens. Actuators B*, **36** (1996), 305. and ref. therein.
28. Arduini, A.; Boldrini, D.; Pochini, A.; Secchi, A.; Ungaro, R. in *Sensors and Microsystems* ed. Di Natale, C.; D'Amico, A., (World Scientific, Singapore, 1996), 25-29.
29. Sansone, F.; Barboso, S.; Casnati, A.; Fabbi, M.; Pochini, A.; Ugozzoli, F.; Ungaro, R., *Eur. J. Org. Chem.* (1998), 897-905
30. Casnati, A.; Fabbi, M.; Pelizzi, N.; Pochini, A.; Sansone, F.; Ungaro, R.; Di Modugno, E.; Tarzia, G., *Bioorg.& Med. Chem. Lett.* **6** (1996), 2699-2704
31. (a) Atwood,J.L.; Koutsantonis, G.A.; Raston, C.L., *Nature,* **368** (1994), 229-231. (b) Suzuki, T.; Nakashima, K.; Shinkai, S., *Chem. Lett.* (1994), 699-702.
32. Ikeda, A.; Yoshimura, M.; Shinkai, S., *Tetrahedron Lett.* **38** (1997), 2107-2110.
33. See *e. g.* Paci, B.; Amoretti, G.; Arduini, A.; Ruani, G.; Shinkai, S.; Suzuki, T.; Ugozzoli, F.; Caciuffo, R., *Phys. Rev. B* **55** (1997), 5566-5569.
34. Tsubaki, K.; Tanaka, K.; Kinoshita, T.; Fuji, K., *Chem. Commun.* (1998), 895-896 and ref. therein.

35. (a) Steed. J.W.; Junk, P.C; Atwood, J.L.; Barnes, M.J.; Raston, C.L.; Burkhalter, R.S., *J. Am. Chem. Soc.* **116** (1994),10346-10347. (b) Atwood, J.L.; Barnes, M.J.; Gardiner, M.G.; Raston, C.L., *Chem. Commun.* (1996), 1449-1450.
36. Hardie, M.J.; Raston, C.L., *Eur. J. Inorg. Chem.* (1999), 195-200.
37. Georghiou, P.E.; Mizyed, S.; Chowdhury, S., *Tetrahedron Lett.* **40** (1999), 611-614.
38. Shinkai, S.; Ikeda, A., *Gazz. Chim. Ital.* **127** (1997), 657-662.
39. Haino, T.; Yanase, M.; Fukazawa, Y., *Angew. Chem. Int. Ed. Engl.* **36** (1997), 259-260.
40. Isaacs, N.S.; Nichols, P.J.; Raston, C.L.; Sandova, C.A.; Young, D.J., *Chem. Commun.* (1997), 1839-1840.
41. Atwood, J.L.; Barbour, L.J.; Nichols, P.J.; Raston, C.L.; Sandoval, C.A., *Chem. Eur. J.* **5** (1999), 990-996.
42. Haino, T.; Yanase, M.; Fukazawa, Y., *Angew. Chem. Int. Ed. Engl.* **37** (1998), 997-998.
43. Yanase, M.; Haino, T.; Fukazawa, Y., *Tetrahedron Lett.* **40** (1999), 2781-2784.
44. Higler, I.; Timmerman, P.; Verboom, W.; Reinhoudt, D.N., *Eur. J. Org. Chem.* (1998), 2689-2702.
45. Van Wageningen, A.M.A.; Timmerman, P.; van Duynhoven, J.P.M.; Verboom, W.; van Veggel, F.C.J.M.; Reinhoudt, D.N., *Chem. Eur. J.* **3** (1997), 639-654.
46. Chapman, R.G.; Sherman, J.C., *Tetrahedron* **53** (1997), 15911-15945.
47. (a) Shimizu, K.D.; Rebek, J.Jr., *Proc. Natl. Acad. Sci. USA* **92** (1995), 12403-12407. (b) Mogck, O.; Böhmer, V.; Vogt, W., *Tetrahedron*, **52** (1996), 8489-8496.
48. Hamann, B.C.; Shimizu, K.D.; Rebek, J.Jr., *Angew. Chem. Int. Ed. Engl.* **35** (1996), 1326-1329.
49. Mecozzi, S.; Rebek, J. Jr., *Chem. Eur. J.* **4**, (1998), 1016-1022.
50. Ikeda, A.; Yoshimura, M.; Tani, F.; Naruta, Y.; Shinkai, S., *Chem. Lett.* (1998), 587-588.
51. Ikeda, A.; Yoshimura, M.; Udzu, H.; Fukuhara, C.; Shinkai, S., *J. Am. Chem. Soc.* **121** (1999), 4296-4297.

CHAPTER 4

CALIXARENES IN SPHERICAL METAL ION RECOGNITION

ALESSANDRO CASNATI AND ROCCO UNGARO
*Dipartimento di Chimica Organica e Industriale dell'Università degli Studi,
Parco Area delle Scienze 17/A, I-43100, Parma, Italy.*

4. 1. Introduction

This chapter deals with the spherical metal ion complexation of calixarene based ligands in solution, with particular attention to the quantitative evaluation of the binding properties, which allows a better comparison with other existing (non calixarene) ligands used for the same purpose. The review will emphasise, whenever possible, the special features of the calixarene ligand and will therefore be critical rather than comprehensive. Recent review articles on this subject have been published and will give a more extensive coverage of the literature up to 1997.[1]

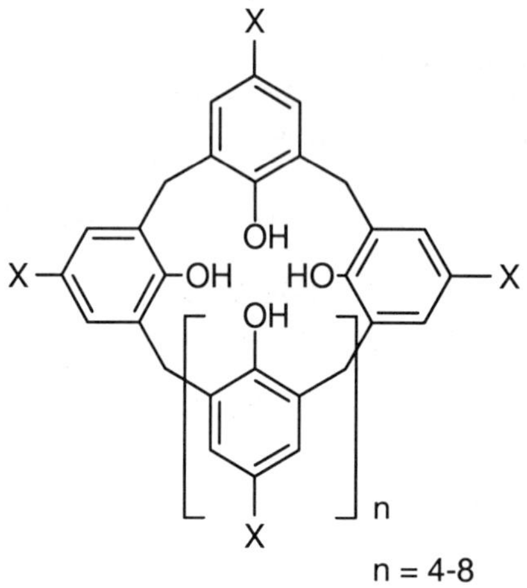

One of the most interesting feature of calixarene-based cation ligands (especially calix[4]arenes) is that the metal ion complexation properties depend not only on the nature of the binding groups attached to the platform, but also on their stereochemical arrangement, which is determined by the calixarene conformation. It is an established fact that the introduction at the lower rim of calix[4]arenes (**1**: n = 4) of alkyl groups larger than ethyl cause the freezing of the ring inversion process, giving four different stereoisomers (*cone, partial cone, 1,3-alternate* and *1,2-alternate*).[1]

Synthetic protocols for the selective functionalisation of the OH groups of calix[4]arenes (*regioselective* alkylation) and for the selective synthesis of certain stereoisomers of tetralkoxy calix[4]arenes (*stereoselective* alkylation) have been developed over the years.[2]

4. 2. Complexation of Alkali Metal Ions

Methoxy derivatives of calix[4]arenes (*e.g.* **2**) have a certain degree of conformational mobility and can adopt one or more structures in solution. Detelier[3] has recently tackled the problem of the kinetics and mechanism of complexation of the alkali metal cations by conformationally mobile tetramethoxy p-*tert*-butyl calix[4]arene **2**, using a combination of several NMR experiments. He was able to show that the Na^+ cation is exclusively complexed by the *cone* conformation, whereas Cs^+ is preferentially complexed by the *partial cone*. These investigations confirm early experimental results showing that simple calix[4]arene ethers **3** in the *cone* conformation preferentially bind the sodium cation.[4] The association constants of these simple calix[4]arene ether podands are, however, rather low (log K < 3 in $CDCl_3$ or CD_3CN) owing to the limited number of binding sites.[1f,i]

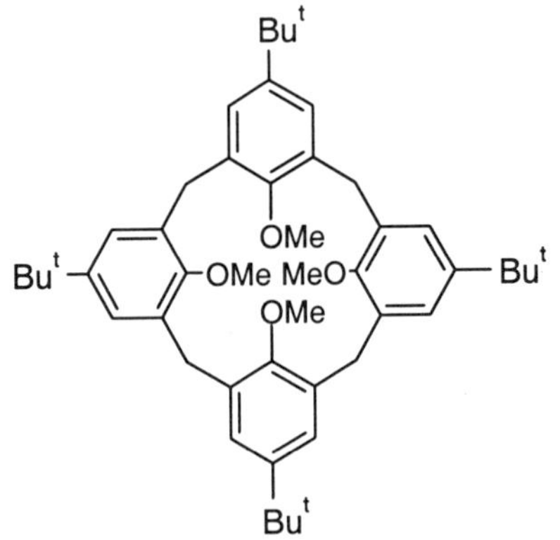

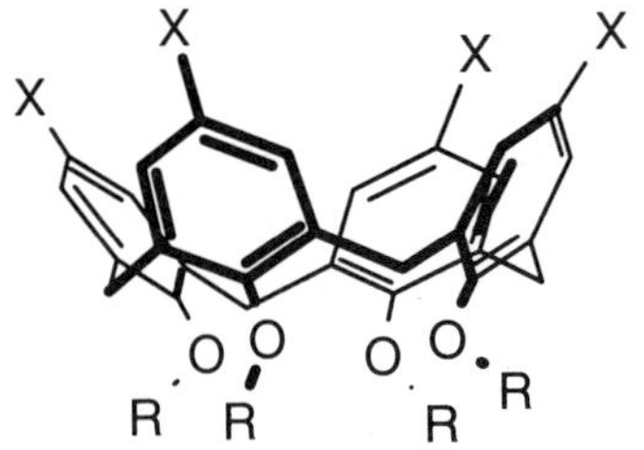

3: X = *t*-Bu; R = $C_2H_4OC_2H_5$

Important information on the conformational preference of calix[4]arene ligands has been obtained by studying 1,3-dimethoxy calix[4]arene crown ethers **4a** and **4b**. Early studies made on the p-*tert*-butyl analogues **4c,d** had shown that they are selective ligands for the potassium and caesium cation respectively, and that, upon potassium complexation the calix adopts a *partial cone* structure.[5] Compounds **4** are conformationally mobile in solution where they exist mainly in the *cone* conformation, which is also shown in the solid state. In the presence of one equivalent of potassium ion, compound **4a** adopts a *1,3-alternate* structure[6] and the same occurs with ligand **4b**[7] in the presence of one equivalent of caesium cation. In the latter case, the X-ray crystal structure of the complex clearly shows this conformational interconversion.[7a] The oxidation of one or two aromatic rings to quinones increases the conformational mobility of calix[4]arene derivatives, which tend to assume a *partial cone* structure in the solid state.[8]

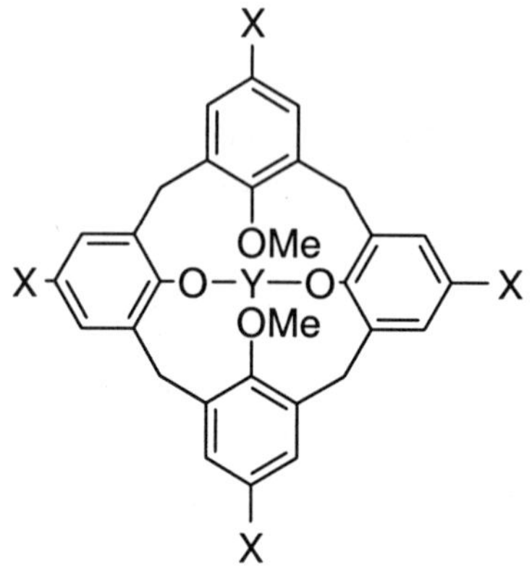

4a: X = H, Y =-(C$_2$H$_4$O)$_3$C$_2$H$_4$-
4b: X = H, Y =-(C$_2$H$_4$O)$_4$C$_2$H$_4$-
4c: X = Bu^t, Y =-(C$_2$H$_4$O)$_3$C$_2$H$_4$-
4d: X = Bu^t, Y =-(C$_2$H$_4$O)$_4$C$_2$H$_4$-

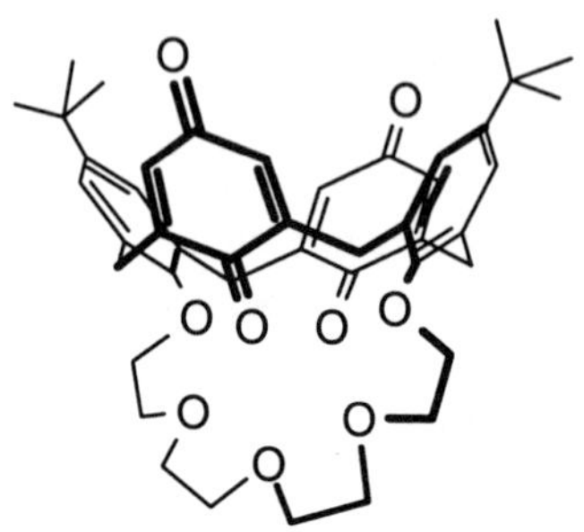

5

Ionophoric calix[4]arene diquinones have been reported by Beer et al.[9] Compound **5** shows strong complexation towards alkali metal salts in methanol (log K > 6 for K$^+$) which is comparable to that of compound **4a** (log K = 6.6±0.2 for K$^+$),[10] although the K$^+$/Na$^+$ selectivity ($\approx$ 10) is lower than that of ligand **4a** (3.1x 10^4).[7a] This can reasonably be explained by the fact that, in the case of ligand **5**, both sodium and potassium are complexed by the ligand in the *cone* conformation, whereas there is clear evidence that sodium is complexed by **4a** in the *cone* conformation and potassium in the *1,3-alternate*.

The information obtained using conformationally mobile ligands has been exploited in the design of more rigid, more efficient, and often more selective calix[4]arene ligands. The early studies on calix[4]arene podands having ether, keto, ester and amide binding groups and blocked in the *cone* conformation have been extensively reviewed.[1i,11] The *cone*, being the most polar conformation of a calix[4]arene, prefers to bind hard metal ions such as Na$^+$ or Ca^{2+}, or trivalent cations such as iron (III) and lanthanides (III). Much greater variation in the selectivity pattern occurs in rigid calixcrowns. By using very efficient synthetic methodologies the conformationally fixed isomers have been obtained in good yields.

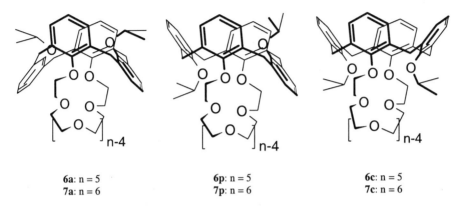

6a: n = 5
7a: n = 6

6p: n = 5
7p: n = 6

6c: n = 5
7c: n = 6

Although all calixcrowns-5 (**4a,c** and **6a,p,c**) are selective for potassium,[6] and all calixcrowns-6 (**4b,d** and **7a,p,c**) for caesium,[5] the efficiency and selectivity of complexation are strongly dependent on the calixarene conformation.

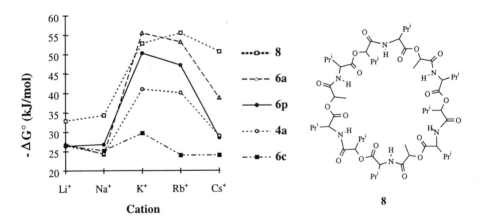

Figure 4.1. Free energies of association (-ΔG°) of calixcrown-5 and valinomycin with alkali picrates in CHCl$_3$ at 25° C.

K$^+$/Na$^+$ selectivity for crowns-5 (Fig. 4.1) and Cs$^+$/Na$^+$ selectivity for crowns-6 decrease in the order *1,3-alternate > partial cone > mobile > cone*, emphasising that the preorganisation of the ligand in the conformation preferred for complexation is a crucial point for obtaining ion selectivity. In particular, the calixcrown-5 (**6**) in the *1,3-alternate* structure shows the highest K^+/Na^+ selectivity ($\Delta\Delta G° = -31.2$ kJ/mol in $CHCl_3$) ever observed for a synthetic ionophore, higher even than valinomycin (**8**), the best potassium selective natural ionophore known. In supported liquid membranes (SLMs),[12] the transport of potassium salts is always dependent on the diffusion coefficient of the complex through the membrane, the only exception being for the derivative (**6p**), in the *partial cone* structure, where the rate limiting step is the release of the salt in the receiving phase. Both in transport (SLM) and in detection with ion selective field-effect transistors (ISFETs) the derivative in the *1,3-alternate* structure (**6a**) shows higher selectivity than valinomycin (**8**).[6]

Also for calixcrowns-6 the stereoisomer (**7a**), fixed in the *1,3-alternate* structure (Fig. 4.2), shows the highest Cs^+/Na^+ selectivity ever observed ($\Delta\Delta G° = -20.2$ kJ/mol in $CHCl_3$).[7] This has allowed us to study the removal of the long-lived radionuclide ^{137}Cs from nuclear waste, where sodium nitrate and nitric acid are present at a much higher concentration than caesium nitrate ($[NaNO_3] = 4M$, $[CsNO_3] = 0.001M$), $[HNO_3] = 1M$).[13]

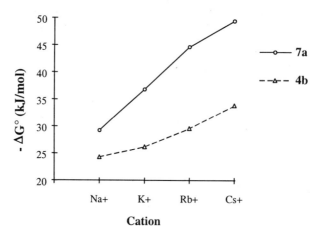

Figure 4.2. Free energies of association ($-\Delta G°$) of calixcrown-6 with alkali picrates in $CHCl_3$ at 25° C.

The use of **7a** in the extraction process or in the transport through supported liquid membranes (SLMs) allows for the recovery of more than 98% of the caesium cation present in solution, making this derivative extremely attractive for the declassification of nuclear wastes.[7b] Ligand **7a** has also been used for the selective detection of caesium in ISEs and ISFETs with very high selectivity and detection limits.[14] We anchored calix[4]arene-crown-5 and -crown-6 derivatives on silica-gel *via* hydrosilanisation and we were able to separate by chromatography potassium or caesium from other alkali metal ions with high efficiency.[15]

Recent Molecular Modeling and X-Ray studies[16] suggested that the incorporation of one or two aromatic units in the crown ether loop in calix[4]arene crown-5 and crown-6 in *1,3-alternate* conformation (*e.g.* **9** and **10**) could improve their K^+/Na^+ or Cs^+/Na^+ selectivity respectively. This could be due to the "stiffening" caused by the benzo ring(s) on the polyether bridge, which cannot wrap easily around the small sodium ion, whereas potassium and caesium enjoy a better size matching and are strongly complexed. Complexation and transport data seem to indicate a better Cs^+/Na^+ selectivity for benzo-calixcrowns in *1,3-alternate* conformation.[16b,17] This effect has been exploited in radioactive waste treatment. The synthesis of calix[4]arene-dibenzo- (**9**) and -monobenzo-crown-6 (**10**) has been performed and both ligands tested for their extraction properties.[18] The presence of benzo units in the crown also ensures an additional position where to introduce long and branched alkyl chains, which greatly increase the solubility of these ligands in the diluents such as NPHE (o-nitrophenylhexyl ether) or THP (hydrogenated tetrapropylene), often used industrially in liquid-liquid extraction.

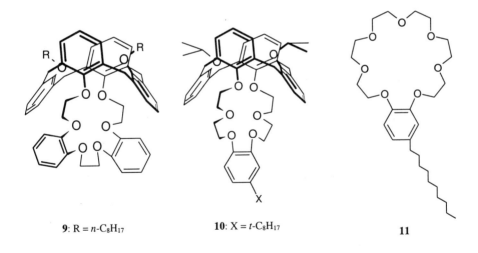

9: R = *n*-C$_8$H$_{17}$ **10**: X = *t*-C$_8$H$_{17}$ **11**

Table 4.1. Distribution coefficients (D) of caesium and sodium nitrates and selectivity $\alpha_{Cs/Na}$.

Ligand	D_{Na}	D_{Cs}	$\alpha_{Cs/Na}$
4b	3.0×10^{-3}	0.04	13
7a	$< 10^{-3}$	28.5	> 28500
10	$< 10^{-3}$	34	> 34000
9	$< 10^{-3}$	31	> 31000
11	1.2×10^{-3}	0.3	250

NOTE: Distribution coefficient (D) is the ratio between metal ion concentration in the organic and aqueous phase at equilibrium; Selectivity ($\alpha_{Cs/Na}$) is the ratio between D_{Cs} and D_{Na}. Aqueous solution, [MNO$_3$] = 5×10^{-4} M – [HNO$_3$] = 1 M; Organic solution, 1×10^{-2} M of ligand in NPHE

As can be seen from Table 4.1, the caesium over sodium selectivity $\alpha_{Cs/Na}$ considerably increases passing from the simple calixcrown-6 **7a** to dibenzo-**9** and monobenzo-crown-6 (**10**), the latter ones being the most caesium selective ligands known so far. Interestingly, the caesium distribution coefficients reach a maximum for nitric acid concentration between 1 and 4 M , while they slightly decrease as sodium nitrate concentration increases from 1 to 4 M. These features thus allow for the caesium ion to be easily stripped in deionised water. However, if necessary, D_{Cs} can be strongly increased by adding to the organic phase the lipophilic cobalt bis dicarbollide, which is known to have a synergistic effect on extraction. Hot tests performed on real radioactive waste show that compound **7a** is able to transport caesium cation very efficiently through a SLM, while most of the other metal ions and radionuclides (actinides and fission products) present are transported less than 1% in the receiving phase.

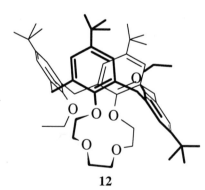

12

A series of sodium selective calix[4]arene crown-4, has been synthesised by Shinkai's group. The selectivity features of these ligands were tested directly on Ion Selective Electrodes (ISEs) where compound **12**, in the *partial cone* structure, shows the highest Na$^+$/K$^+$ selectivity found so far for a synthetic ionophore (log $K^{pot}_{Na/K}$ = -5.3).[19] These data were also confirmed in an independent study.[20]

Similar properties to those previously described for simple 1,3-dialkoxycalix[4]arene crown ethers, including the effect of the benzo unit in the polyether ring,[13a,21] have been obtained with calix[4]arene bis crown ethers in *1,3-alternate* conformation,[22] having two identical crown ether loops. A wider range of novel binding properties could be envisaged in the case of calixbiscrowns having inequivalent crown loops, which could in principle lead to the encapsulation of two different metal ions. Only very recently have these possibilities started to be explored.[23]

Ionophoric biscalix[4]arenes of the type **13** and **14** have recently been reported.[1g,24,25]

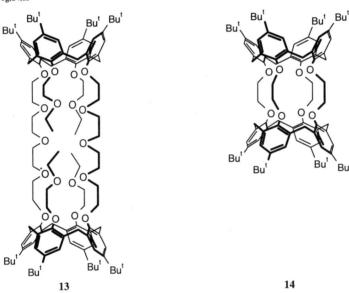

13   14

When the length of the polyether chains linking the two calixarene subunits is sufficiently high as in compound **13**, two binding regions are created, each close to the calixarene oxygen atoms, and the complexed cation (sodium or potassium) oscillates between the two sites. Both intra- and intermolecular exchanges between the two sites occur, and it was possible to distinguish between the two processes by means of ^{1}H NMR experiments performed at different temperatures and concentrations.[24] On the other hand, compound **14** is a very rigid and rather insoluble tubular receptor (*calix[4]tube*) with considerable potassium selectivity. Solid-liquid extraction experiments using alkali metal iodides, show an easy uptake only for KI. The X-Ray crystal structure of the KI complex shows that the cation is encapsulated in the central oxygen atom region, although molecular mechanics calculations suggest that the preferred pathway for potassium uptake involves a prior weak complexation of the cation with the calix[4]arene aromatic rings.[25] This conclusion could have relevance for the construction of biomimetic artificial potassium channels.

4. 3. Complexation of Alkaline-Earth Metal Ions

Although calix[4]arene tetraesters (**3**: R = CH$_2$COOR') and tetraketones (**3**: R = CH$_2$COR') in the *cone* conformation are able to complex alkali metal cations, they are ineffective in binding of alkaline-earth cations.[1i]

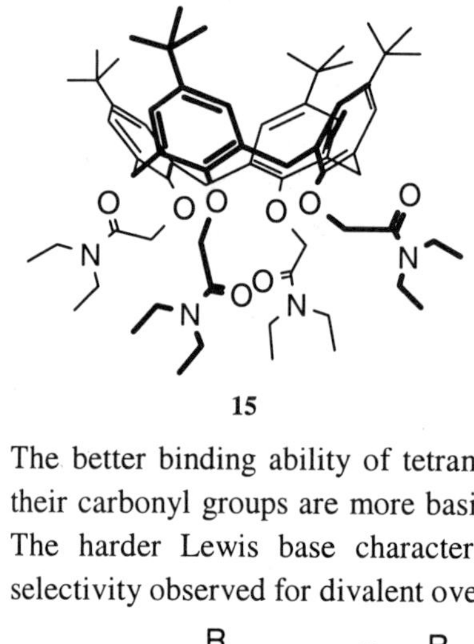

In contrast, calix[4]arene tetramides, with a large variety of different substituents on the nitrogen atoms, are able to complex these cations efficiently. Of the various tertiary amides studied, tetradiethylamide **15**[26] is the most efficient, with association constants at least as high as those of cryptand 221 (Logβ = 9.3). Log β in methanol (≥ 9 for Ca^{2+} and Sr^{2+}) and percentage of extraction (E%) from water to dichloromethane indicate the following selectivity order: Ca^{2+}≈ Sr^{2+} > Ba^{2+} >>Mg^{2+}.[27]

The better binding ability of tetramide derivatives is certainly due to the fact that their carbonyl groups are more basic than those of the ester and ketone derivatives. The harder Lewis base character of the amide moiety also accounts for the selectivity observed for divalent over monovalent cations.

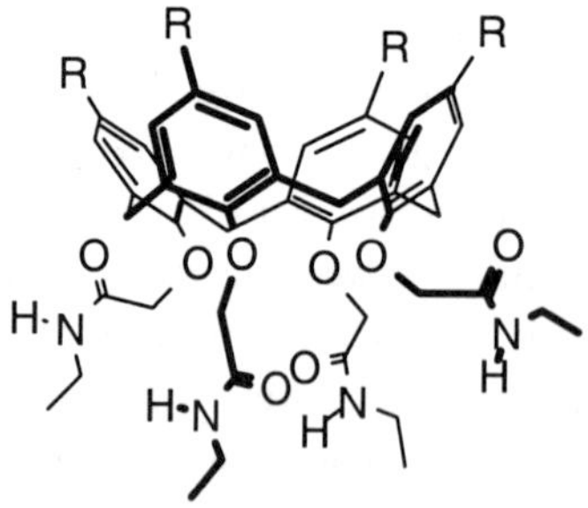

16: R = *t*-Bu

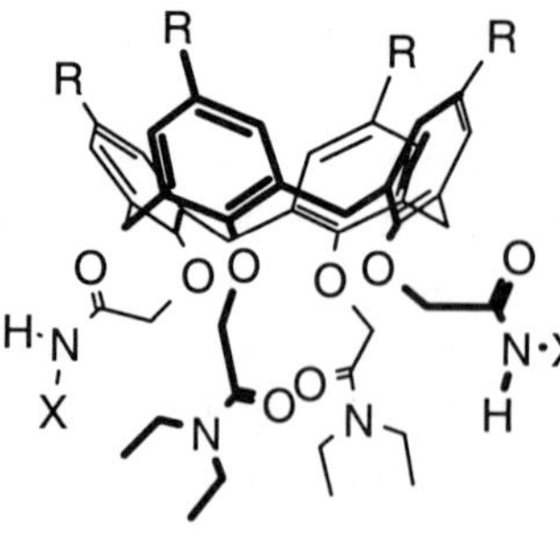

17a: R = *t*-Bu; X = H
17b: R = *t*-Bu; X = n-Bu

The binding ability of tertiary amide groups has also been compared to that of secondary and primary amides.[28] Extraction percentage (E%) is strongly depressed passing from the tetra-tertiary amide **15** to tetra-secondary amide derivative **16**, the latter showing E% very close to zero both for monovalent and divalent cations. The effect of different substitution on amide groups has recently been investigated by comparing the three tetramides **15** and **17a,b** which differ in that they have, respectively, tertiary, secondary and primary amides in a diametrical position at their lower rims. Binding properties both in extraction (CH_2Cl_2) and in complexation (in methanol) indicate substantially decreasing efficiency following the order **15>17a>17b**, with a more pronounced effect on alkalis than on alkaline-earths and with a shifting of selectivity towards the latter. Although this behaviour may in part be due to the possibility of secondary and, even more likely, of primary amides to form intramolecular H-bonding with the neighbouring phenolic oxygens - as has also been seen from ^{1}H NMR studies in $CDCl_3$ - molecular modeling calculations do not support this hypothesis.[28]

Calixquinones have been also used for the complexation of alkaline-earth metal ions. Significant cation complexation has been reported when at least two aromatic nuclei (usually the diametrical ones) were functionalised with efficient binding groups such as esters, amides, crown ether bridges and carboxylic acids. For non-ionizable ligands **18** the change of the phenolic nuclei to quinone, changes significantly the possible co-ordination around the cations, a slight preference for alkaline-earth metals usually being observed.

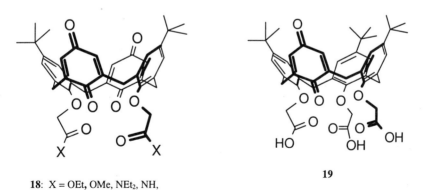

18: X = OEt, OMe, NEt$_2$, NH,
n-Bu, NH$_2$

19

Upon metal ion complexation, the conformational mobility of quinone nuclei in calix[4]arenes is blocked and the *cone* conformation is strongly favoured, allowing the quinone carbonyls to enter the co-ordination sphere. The possibility of interaction between the complexed ion and the quinone carbonyl groups strongly influences the electrochemistry of these nuclei.[9,29,30] In all cation complexes the potentials of reduction are significantly shifted anodically, allowing these derivatives to be used as possible amperometric sensing devices. This effect is particularly noticeable for alkaline-earth metal ions, since they have a higher charge density than alkali ions. Calixquinone **19** bearing ionisable carboxylic acid groups[31] was studied in water, showing a selective response towards Ca^{2+} even in the presence of a large excess of interfering Na^+ ion.

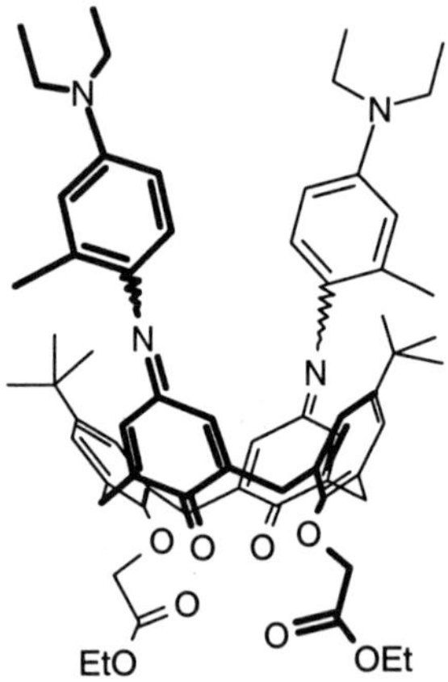

20 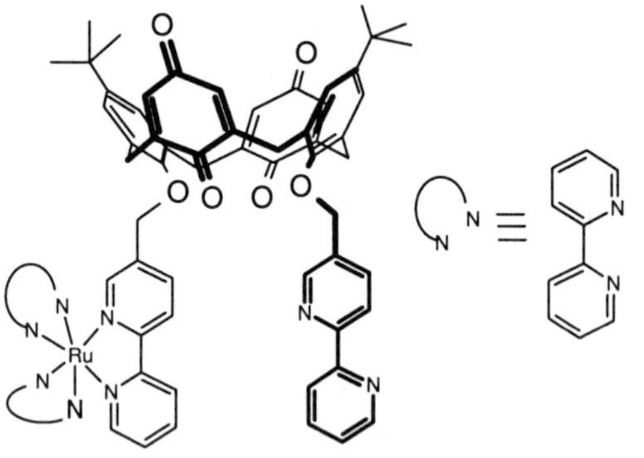   **21**

The use of indoaniline derivative **20**,[32] thanks to the intimate interaction of the quinone carbonyl group of the chromophore with the complexed cation, gave rise to a chromogenic receptor which showed considerable selectivity in the detection of calcium ion. Very recently[33] a luminescent ruthenium (II) tris(2,2'-bipyridyl) fragment was appended on a calix-diquinone structure, obtaining a photoionic detector **21** of alkali and alkaline-earth metal ions. The intramolecular electron transfer which takes place between the quinone and $Ru(bipy)_3$, and which causes a quenching of luminescence, is inhibited by the complexation of cations because of a conformation rearrangement of the redox active units. This allows for the detection of the complexation event, by an increase in luminescence yield and lifetime.

As predicted by the hard and soft acid base (HSAB) principle,[34] an efficient complexation of hard cations such as the alkaline-earths or the tri- and tetravalent metal ions requires the use of hard donor groups such as carboxylates.[35] This is also supported by several X-ray structures of calcium binding proteins which indicate that one to four carboxylate groups are always co-ordinated to the metal ion as either mono- or bidentate ligands.[36] Also with the purpose to complex di- and trivalent metal ions, several examples of calix[4]arenes bearing four to one hydroxycarbonyl methoxy groups at the lower rim were reported.

The first example is the tetracid of p-*tert*-butylcalix[4]arene **22**, which shows a high efficiency in the extraction of alkali-earth metal ions.[37] Several complexation studies have been performed to evaluate the binding properties of tetracid **22** towards alkaline and alkaline-earth,[38a,b] lanthanide,[38c] and actinide[38d] cations. Selectivity among alkaline-earth cations is for calcium ion but strontium is also efficiently complexed.

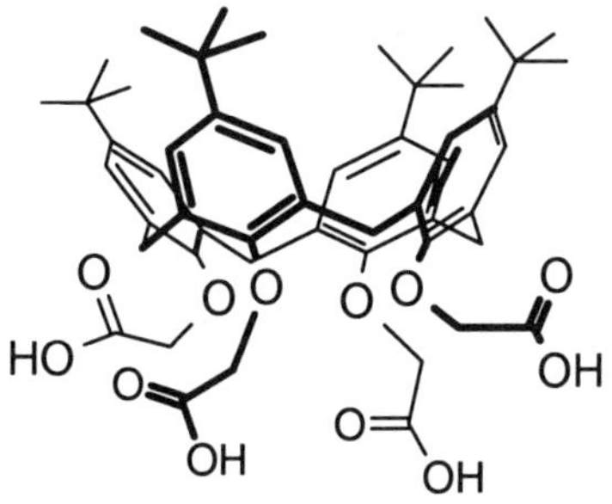

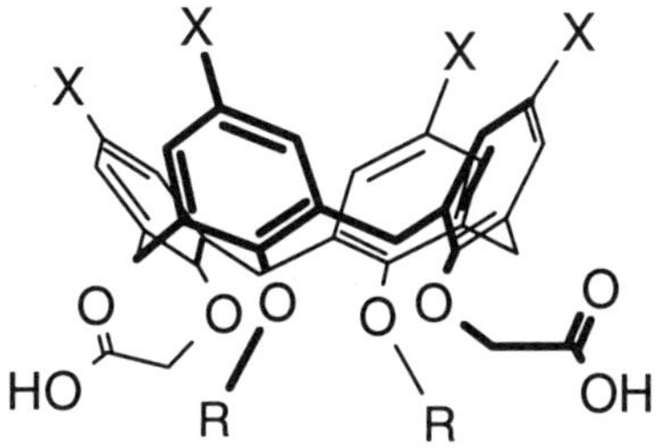

23a: R = CH$_3$, X = t-Bu
23b: R = H, X = t-Bu
23c: R = CH$_2$CONEt$_2$, X = t-Bu
23d: R = CH$_2$CONEt$_2$, X = H

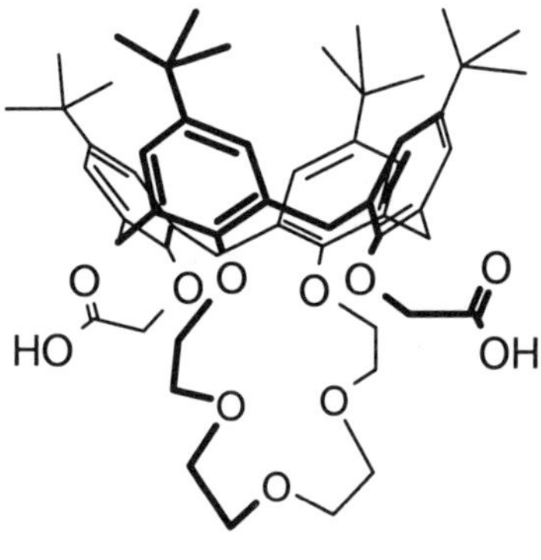

Calix[4]arenes (**23**) diametrically substituted at the lower rim with two hydroxycarbonyl methoxy groups have also been prepared,[37a,39] but for several of them[39d-h] the acidic groups were only used for further functionalisation. 1,3-Dihydroxy- (**23b**), 1,3-dimethoxy- (**23a**) and crown-5 (**24**) diacid derivatives show high efficiency in extraction of alkali-earth cations from water to dichloromethane. The crown-5 derivative **24**, shows the highest efficiency and selectivity towards the calcium ion, probably because of the presence of the carboxylic acid donor groups and of the size of the polyether bridge. More recently, Shinkai et al.[39b] reported on the extraction properties of the diamide-diacid derivative of p-*tert*-butylcalix[4]arene (**23c**). This compound, combining two hard amide and carboxylic groups, shows a high selectivity for calcium, strontium being the second best extracted cation within alkali-earth metals. We have recently found that compound **23d**, which derives from calix[4]arene, shows an even higher Sr^{2+}/Na^+ selectivity.[40] The binding properties of diamide-diacids **23c-d** in methanol indicate that both ligands present even higher efficiency in the complexation of divalent alkaline-earths than tetracid **22**.

4. 4. Complexation of Lanthanide Ions

Calixarenes have been extensively used as hosts for the complexation of lanthanide ions. Much work has been done on simple non-functionalised calixarenes where the phenolic OH groups are ionised and act as ligands to give unusual structures. The most interesting have been reviewed by Ugozzoli in this book, whereas most of the work has been extensively covered in recent review articles.[1i,41,42] Since the early report by our group of the ability of calix[4]arene tetramide **15** to form luminescent complexes with lanthanide ions (especially Tb^{3+}) in water (Fig. 4.3.a),[43] several groups have tried to synthesise new ligands with improved photophysical properties. The main feature of tetramide **15** is its ability to encapsulate lanthanide ions giving rise to an efficient *antenna effect* (Fig. 4.3.b). Only one water molecule has been found to be co-ordinated around the **15**.Ln^{3+} complex, while in cryptands 3-4 water molecules usually surround the cation.[44]

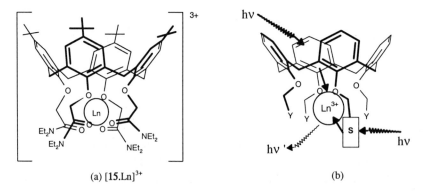

Figure 4.3. a) Complex of ligand **15** with a lanthanide (III) metal ion and b) antenna effect in calix[4]arene lanthanide complexes: S = sensitiser; Y = binding group.

A weak point for this ligand is its low molar absorbitivity and the poor quantum yield observed for the Eu^{3+} complexes which was attributed to the deactivation of the ligand excited states *via* ligand to metal charge transfer states. One approach followed was to keep constant the *cone* conformation of the ligand and to attach better chromophores. Calix[4]arenes bearing 2,2'-bipyridine moieties at the lower rim have been synthesised.

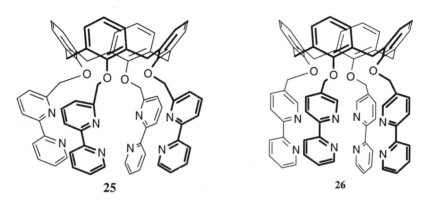

Compound **25** has the four bipyridine units attached to the calixarene through the C^6 carbon (ortho position to the pyridine nitrogen),[45] whereas compound **26** has the same units linked *via* the C^5 carbon (meta position to the pyridine nitrogen).[46] Both compounds form 1:1 complexes with Eu^{3+} and Tb^{3+} characterised by high luminescence intensities which are mainly a consequence of the high molar absorbitivity of the ligands. Although the quantum yield of the luminescence is not very high for either ligands it is interesting to note that it is higher for terbium than for europium in the case of ligand **25**, whereas the reverse is true in the case of ligand **26**. The slight better photophysical properties shown by the lanthanide complexes of ligand **26** compared with **25** have been attributed to a better wrapping of the four bipyridine units around the lanthanide ion offered by C^5 linkage, whereas the bipyridine units attached through the C^6 could be affected by some steric hindrance.[45,46] Molecular Modeling results seems to confirm this hypothesis.[46,47] One of the drawbacks of the ligands containing four 2,2'-bipyridine units is the low stability of their lanthanide complexes (Log K ~ 5 in CH_3CN)[44] owing to the poor complexation ability of the soft chelating units.

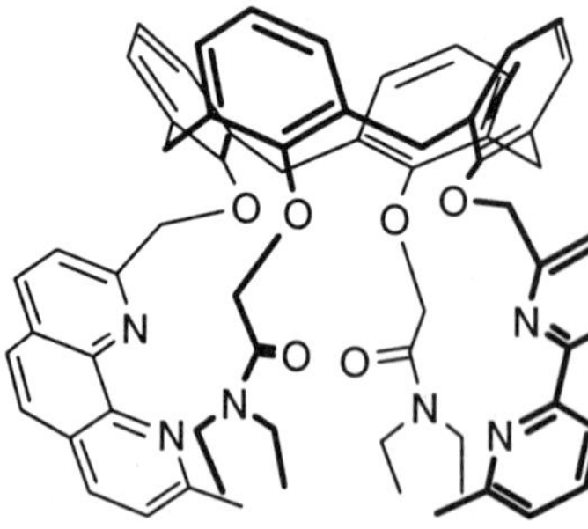

27

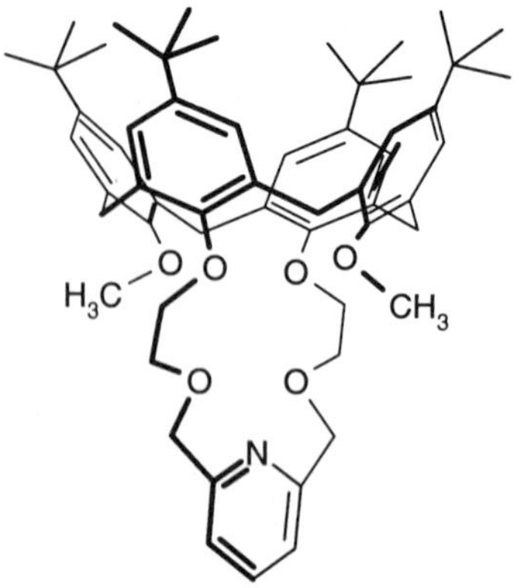

28

The synthesis of ligands bearing a mixed type of chelating groups such as amide-bipyridine or amide-phenantroline (*e.g.* **27**)[45] or having a macrobicyclic structure, such as **28**,[48] makes for a good compromise between stability and luminescence properties of the corresponding complexes with lanthanide ions. Shinkai *et al.* have used calix[4]arene triamides in the *cone* conformation having different phenacyl chromophores on the remaining pendant arm.[49] One of these derivatives, **29**, bears an (o-boronylphenyl)methyl group which acts as a sugar binding site. The phenyl group of the phenacyl moiety can act as a sensitiser for lanthanide ions, but the intramolecular amine is known to quench its fluorescence as well. However, upon the addition of a saccharide (Fig. 4.4), the interaction between the nitrogen and the boron atom becomes stronger owing to the binding of boronic acid to the sugar, resulting in a decreased fluorescence quenching. It has been established that, on the addition of D-fructose, the luminescence intensity of **29**.Eu^{3+} is enhanced approximately 9 times, whereas that of **29**.Tb^{3+} was approximately double. This has led to the conclusion that the Eu^{3+} ion is mainly sensitised by the phenacyl group and the Tb^{3+} ion mainly by the calix[4]arene aromatic rings.[50]

In most cases discussed above the reported luminescence measurements were performed in acetonitrile and not in a competitive solvent like methanol or water. Efficiency in water is usually substantially lower owing to the O-H deactivating effect. Moreover, the ligands are neutral and the complexes are highly charged. One way of overcoming these problem is to use calixarene ligands having carboxylate binding groups, which are also known to effectively shield the lanthanide cations in water or methanol.

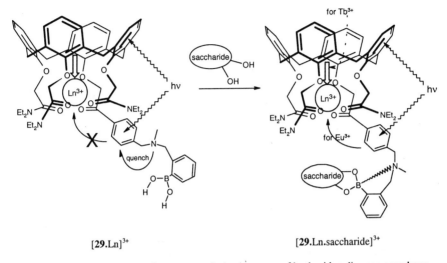

Figure 4.4. Saccharide control of energy-transfer luminescence of lanthanide calixarene complexes.

Calix[4]arene tri- and tetracarboxylic acids in the *cone* conformation are known to form stable 1:1 complexes with lanthanide ions at basic pH (Logβ_{MeOH} = 23-25). The topic has recently been reviewed.[42] Reinhoudt's group has extensively exploited calix[4]arene triacid derivatives bearing different types of chromophores on the fourth pendant arm to prepare and study neutral luminescent lanthanide complexes.[51-53]

31: X = NHCH$_2$CON(CH$_2$OH)$_3$

When a triphenylene antenna chromophore was attached (30), a strong sensitising ability toward Eu^{3+} and Tb^{3+} was obtained, allowing for the excitation of the lanthanide ions with wavelengths extending to 350 nm.[52] A water soluble calix[4]arene triacid-monoamide 31 with a chrysene moiety as sensitiser shows strong lanthanide emission for Eu^{3+} with an excitation maximum at $\lambda = 363$ nm.[53]

4. 5. Complexation of Actinides

Inter- and intragroup separations of lanthanides and actinides are among the most interesting goals to pursue in the management of nuclear waste. Neutral organophosphorus compounds are among the most useful extractants for these purposes and are currently used in various industrial processes.

Plutonium and uranium separation in PUREX process is achieved with tributylphosphate, while CMPOs which combine a phosphine oxide and a carbamoyl group, are used in TRUEX process for the decategorisation of liquid waste. Obviously, increasing attention is now being devoted to the introduction of phosphorus-containing binding groups at the lower or upper rim of calixarenes, in order to study the effect of preorganisation on the extraction efficiency and selectivity of lanthanides and actinides.

32: n = 4, 6, 8; m = 1, 2

McKervey et al.[54,55] have synthesised several calix[4]-, -[6]- and -[8]arene ligands bearing phosphine oxide groups at the lower rim (32) while we[56] and others[57] have prepared calix[4]- (33) and calix[6]arenes (34) with mixed crown-ether/phosphine oxide or amide/phosphine oxide groups. In general most of these derivatives show higher extraction percentages of Th (IV) and Eu (III) from water to dichloromethane than TOPO (trioctylphosphine oxide) or CMPO, especially at low ligand concentration (10^{-3}-10^{-2} M), indicating that co-operation is effective among the binding groups on the calixarene.

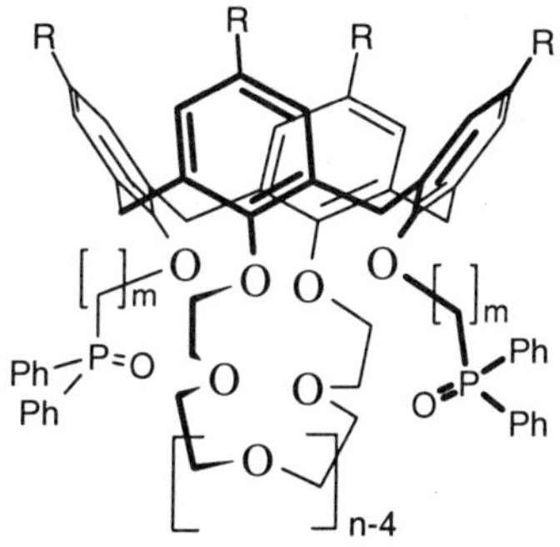

33: n = 5, 6; m = 1, 4

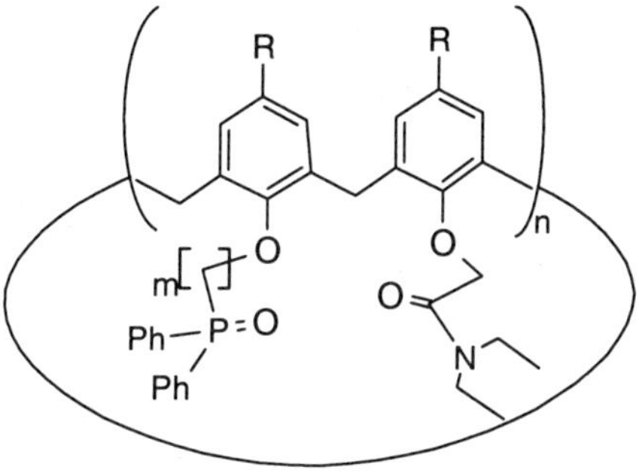

34: n = 2, 3; m = 1, 4

An even more evident co-operative effect has been obtained by the introduction of CMPO functions at the upper (*e.g.* **35**)[58,59] and lower (**36**)[60] rim of calix[4]arenes. All these derivatives show high distribution coefficients from water to NPHE of actinide and lanthanide ions also at high nitric acid concentration. Transport studies with ligands **35** show high permeabilities of Pu and Am through SLMs where the source phase is 1M in HNO_3, 4M in $NaNO_3$. Under these conditions, Pu can be quantitatively removed from simulated nuclear waste after 6 hours. To obtain a comparable result with CMPO, the latter should be used concentrated at least 10-fold more.

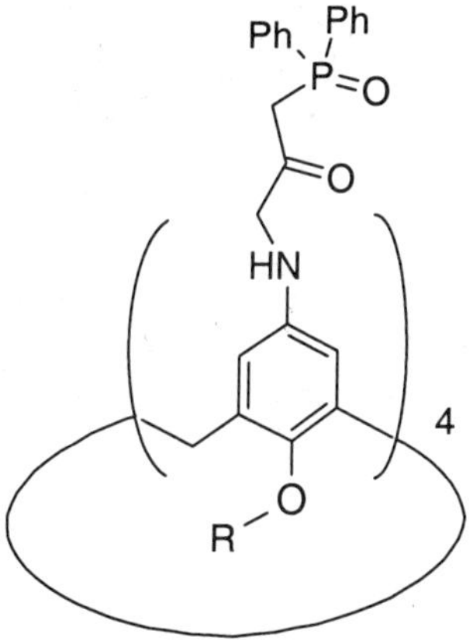

35

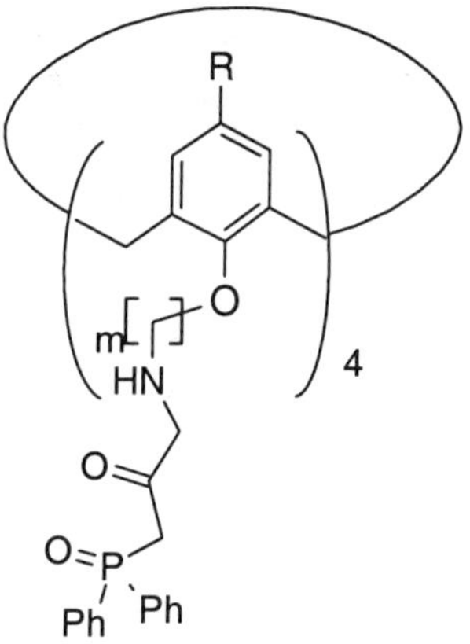

36 m = 2, 3, 4

Compound **35** (R = n-C_5H_{11}) also exhibits considerable selectivity based mainly on the size of the cations, the actinides (and light lanthanides) being better extracted than the heavy lanthanides.

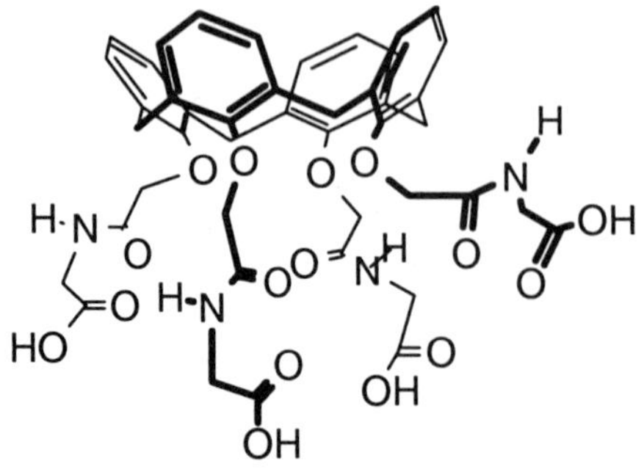

Also ionisable ligands have been used for this purpose. Reinhoudt *et al.* found by Molecular Modeling, and then confirmed by experimental results,[61] that calix[4]arene bearing multiple glycine units (*e.g.* **37**), having an increased number of potential hard donor groups, have slightly better complexing properties towards La^{3+} used as non-radioactive model of Ac^{3+}.

However, the use of all these carboxylic acid derivatives in radioactive waste treatment is restricted by the severe drawback that all these ligands show complexation abilities only at pH higher than 3-4.

Hydroxamic acid chelating chains, which are effective for Pu (IV) complexation, have been introduced at the lower rim of calix[4]arenes by us[62] and others.[63] Although compounds **38** and **39** can efficiently extract Th (IV) their higher affinity for Fe (III) and their hydrolytic lability under acid pH prevent their use on radioactive waste.[63]

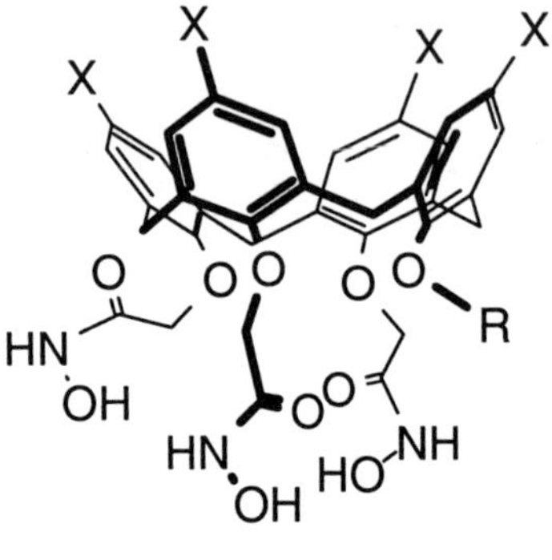

38a: R = CH₃, X = H, *t*-C₄H₉
38b: R = CH₂CONHOH, X = H, *t*-C₄H₉

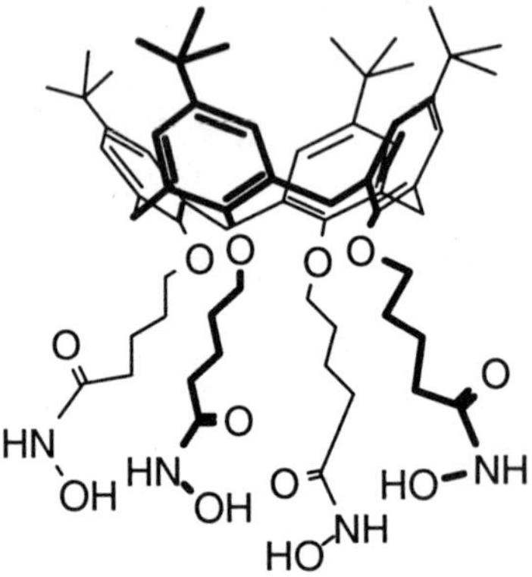

It has recently been reported that both calix[4]arene tetra- and calix[6]arene hexa-carboxylic acid derivatives are able to bind actinium-225 (^{225}Ac) but not biologically important cations such as Na⁺, K⁺, Mg²⁺, Ca²⁺, and Zn²⁺.[38d] Since ^{225}Ac is an α-emitter with $t_{1/2}$ = 10 days, this finding opens up new possibilities in the use of its calixarene complexes in radioimmunotherapy.

4. 6. Acknowledgements

This work was supported by M.U.R.S.T (Ministero dell'Università e della Ricerca Scientifica e Tecnologica), Progetto "Dispositivi Supramolecolari", and by CNR.

4. 7. References

1. a) Gutsche C.D., *Calixarenes Revisited*, ed. Stoddart J. F. (The Royal Society of Chemistry, Cambridge, 1998). b) Casnati A., *Gazz. Chim. Ital.* **127** (1997), 637–649. c) Ungaro R., Arduini A., Casnati A., Ori O., Pochini A., Ugozzoli F., in *Computational Approaches in Supramolecular Chemistry*, ed. Wipff G. (Kluwer Academic Publishers, Dordrecht, 1994), Nato ASI Series C 426, 277–299. d) Arduini A., Casnati A., Dalcanale E., Pochini A., Ugozzoli F., Ungaro R. in *Supramolecular Chemistry: Where it is Where it is going*, eds. Dalcanale E. and Ungaro R. (Kluwer Academic Publishers, Dordrecht, 1999), Nato ASI Series C527, 67–94. e) Ungaro R., Arduini A., Casnati A., Pochini A., Ugozzoli F., *Pure & Appl. Chem.* **68** (1996), 1213–1218. f) De Namor A. F. D., Cleverley R. M., Zapata Ormachea M. L., *Chem. Rev.* **98** (1998), 2495–2525. g) Ikeda A., Shinkai S., *Chem. Rev.* **97** (1997), 1713–1734. h) Arnaud-Neu F., Schwing-Weill M. J., *Synthetic Metals* **90** (1997), 157–164. i) McKervey, M. A., Schwing-Weill M.-J., Arnaud-Neu F., in *Comprehensive Supramolecular Chemistry*, ed. Gokel G. W., (Pergamon, Oxford, 1996), Vol. 1, 537–603.
2. a) Groenen L. C. *et al*, *Tetrahedron Lett.* **32** (1991), 2675–2678. b) Arduini A., Casnati, A. (1996) "Calixarenes" in *Macrocyclic Synthesis: a Practical Approach*, ed. Parker D. (Oxford University Press, Oxford, 1996), 145–173.
3. Meier U. C., Detellier C., *J. Phys. Chem. A* **102** (1998), 1888–1893.
4. Chang S. K., Cho I., *J. Chem Soc., Perkin Trans 1* (1986), 211–214.
5. a) Dijkstra P. J. *et al*, *J. Am. Chem. Soc.* **111** (1989), 7567–7575. b) Ghidini E., Ugozzoli F., Ungaro R., Harkema S., El-Fadl A. A., Reinhoudt D. N., *J. Am. Chem. Soc.* **112** (1990), 6979–6985.
6. Casnati A. *et al*, *Chem. Eur. J.* **2** (1996), 436–445.
7. a) Ungaro R. *et al*, *Angew. Chem. Int. Ed. Eng.* **33** (1994), 1506–1509. b) Casnati A. *et al*, *J. Am. Chem. Soc.* **117** (1995), 2767–2777.
8. Casnati A. *et al*, *Recl. Trav. Chim. Pays-Bas* **112** (1993), 384–392.
9. Beer P. D. *et al*, *Inorg. Chem.* **36** (1997), 5880–5899.
10. Arnaud-Neu *et al*, *Gazz. Chim. Ital.* **127** (1997), 693–697.

11. Andreetti G. D, Ugozzoli F., Ungaro R., Pochini A., in *Inclusion Compounds*, eds. Atwood J. L., Davies J. E. D. and MacNicol D. D. (Oxford Science Publications, Oxford, 1991) Vol. 4, 64–125.
12. Visser H. C., Reinhoudt D. N., de Jong F., *Chem. Soc. Rev.* **23** (1994), 75–81.
13. a) Hill C. *et al*, *J. Incl. Phenom. Molecul. Recogn. Chem.* **19** (1995), 399–408. b) Dozol J. F., Rouquette H., Ungaro R., Casnati A., Patent FR 93-4566 930419, WO 94-FR432 940418, *C.A.* **122** (1995) 239730.
14. a) Lugtenberg, R. J. W., Brzozka Z., Casnati A., Ungaro R., Engbersen J. F. J., Reinhoudt D. N., *Anal. Chim. Acta* **310** (1995), 263–267. b) Bocchi C., Careri, M., Casnati A., Mori G., *Anal. Chem.* **67** (1995), 4234–4238.
15. Arena G., Casnati A., Contino A., Mirone L., Sciotto D., Ungaro R., *J. Chem. Soc., Chem. Commun.* (1996), 2277–2278.
16. a) Lamare V. *et al*, *Sep. Sci. and Techn.* **32** (1997), 175–191. b) Lamare V., Dozol J.-F., Ugozzoli F., Casnati A., Ungaro R., *Eur. J. Org. Chem.* (1998), 1559–1568.
17. Kim J. S. *et al*, *J. Chem. Soc., Perkin Trans. 2* (1999), 837–846.
18. Dozol J. F., Lamare V., Simon N., Ungaro R., Casnati A., *Proceedings ACS Symposium "Calixarene Molecules for Separation"*, **217** ACS Meeting, Anaheim, Ca , March 21-25 (1999), in press.
19. Yamamoto H., Shinkai S., *Chem. Lett.* (1994), 1115–1118.
20. Bakker E., *Anal. Chem.* **69** (1997), 1061–1069.
21. Kim S. J., Shu I. H., Kim K. J., Cho M. H., *J. Chem. Soc. Perkin Trans. 1* (1998), 2307–2311.
22. Lamare V. *et al*, *J. Chem. Soc., Perkin Trans. 2* (1999), 271–284. Asfari Z., Lamare V., Dozol J.-F., Vicens J., *Tetrahedron Lett.* **40** (1999), 691–694. Thuery P., Nierlich M., Lamare V., Dozol J.-F. Asfari Z., Vicens J., *Supramol. Chem.* **8** (1997) 319–332. Asfari Z. *et al*, *New J. Chem.* **20** (1997), 1183–1194. Arnaud-Neu F., Asfari Z., Souley B., Vicens J., *New J. Chem.* **20** (1996), 453–463.
23. Aeungmaitrepirom W., Asfari Z., Vicens J., *Tetrahedron Lett.* **38** (1997), 1907–1910. Asfari Z. *et al*, *J. Incl. Phenom. Macroc. Chem.* **33** (1999), 251–262.
24. Ohseto F., Shinkai S., *J. Chem. Soc., Perkin Trans. 2* (1995), 1103–1109.
25. Schmitt P., Beer P. D., Drew M. G. B., Sheen P. D., *Angew. Chem., Int. Ed. Engl.* **36** (1997), 1840–1842.
26. Arduini A. *et al*, *J. Incl. Phenom. Molecul. Recogn. Chem.* **6** (1988), 119–134.
27. Arnaud-Neu F., Schwing-Weill M.-J., Ziat K., Cremin S., Harris S. J., McKervey M. A., *New J. Chem.* **15** (1991), 33–37. Arnaud-Neu F. *et al*, *J. Chem. Soc., Perkin Trans. 2* (1995), 453–461.

28. Arnaud-Neu F. *et al, J. Chem. Soc., Perkin Trans. 2* (1999), 1727–1738.
29. Chung, T. D., Choi D., Kang S. K., Lee S. K., Chang S.-K., Kim H., *J. Electroanal. Chem.* **396** (1995), 431–439.
30. Beer P. D., Gale P. A., Chen G. Z., *J. Chem. Soc., Dalton Trans.* (1999), 1897–1909.
31. Chung T. D., Kang S. K., Kim H., Kim J. R., Oh W.S., Chang S. K. *Chem. Lett.* (1998), 1225–1226.
32. Kubo Y., Tokita S., Kojima Y., Osano Y. T., Matsuzaki T., *J.Org. Chem.* **61** (1996), 3758–3765.
33. Harriman A., Hissler M., Jost P., Wipff G., Ziessel R., *J. Am. Chem. Soc.* **121** (1999), 14–27.
34. Pearson R. G., *J. Am. Chem. Soc.* **85** (1963), 3533–3539.
35. Arduini A., Casnati A., Pochini A., Ungaro R., *Curr. Opin. Chem. Biol.* **1** (1997), 467–474.
36. Kaufman Katz A., Glusker J. P., Beebe S. A., Bock C. W., *J. Am. Chem. Soc.* **118** (1996), 5752–5763.
37. a) Ungaro R., Pochini A., Andreetti G. D., *J. Incl. Phenom. Molecul. Recogn. Chem.* **2** (1984), 199–206. b) Arduini A., Pochini A., Reverberi S., Ungaro R., *J. Chem. Soc., Chem. Commun.* (1984), 981–984.
38. a) Schwing-Weill M.-J., Arnaud-Neu F., McKervey M. A., *J. Phys. Org. Chem.* **5** (1992), 496–501. b) Arnaud F., *et al, Inorg. Chem.* **32** (1993), 2644–2650. c) Arnaud-Neu F. *et al*, *J. Chem. Soc., Dalton Trans.* (1997), 329–334. d) Chen X., Ji M., Fisher D. R., Wai C. M., *Chem. Commun.* (1998), 377–378.
39. a) Ungaro R., Pochini A., in *Calixarenes: a Versatile Class of Macrocyclic Compounds*, eds. Vicens J., Böhmer V. (Kluwer Academic Publishers, Dordrecht, 1990), 127–147. b) Ogata M., Fujimoto K., Shinkai S., *J. Am. Chem. Soc.* **116** (1994), 4505–4506. c) Montavon G., Duplatre G., Barakat N., Burgard M., Asfari Z., Vicens J., *J. Incl. Phenom. Molecul. Recogn. Chem.* **27** (1997), 155–168. d) Collins E. M., McKervey M. A., *J. Chem. Soc., Perkin Trans. 1* (1989), 372–374. e) Rudkevich D. M., Stevens T. W., Verboom W., Reinhoudt D. N., *Recl. Trav. Chim. Pays-Bas* **110** (1991), 294–298. f) Murakami H., Shinkai S., *Tetrahedron Lett.* **34** (1993), 4237–4240. g) Zhong Z.-L., Tang C.-P., Wu C.-Y., Chen Y.-Y., *J. Chem. Soc., Chem. Commun.* (1995), 1737–1738.
40. Barboso S., Casnati A., Arnaud-Neu F., Schwing-Weill M.-J., Ungaro R., (1999) submitted for publication.
41. Roundhill D. M., *Prog. Inorg. Chem.* **43** (1995), 533–592.
42. Arnaud-Neu F., *Chem. Soc. Rev.* **23** (1994), 235–241.
43. Sabbatini N. *et al*, *J. Chem. Soc., Chem. Commun.* (1990), 878–879.
44. Sabbatini F., Guardigli M., Manet I., in *Handbook on the Physics and*

Chemistry of Rare Earths, eds. Gschneider K. A., Eyring L. (Elsevier Science, London, 1996), chpt. 154, 19–117. Sabbatini F., Guardigli M., Manet I., *Advances in Photochemistry* **23**, ed. Neckers D. (1997), 213–278.
45. Casnati A. *et al, J. Chem. Soc., Perkin Trans.* 2 (1996), 395–399.
46. Ulrich G. *et al, Chem. Eur. J.* **3** (1997), 1815–1822.
47. Fraternali F., Wipff G., *J. Phys. Org. Chem.* **10** (1997), 293–304. van Veggel F. C. J. M., *J. Phys. Chem. A* **101** (1997), 2755–2765.
48. Sabbatini N. *et al, Inorg. Chim. Acta.* **252** (1996), 19–24.
49. Sato N., Shinkai S., *J. Chem. Soc., Perkin Trans.* 2 (1993), 621–624. Matsumoto H., Shinkai S., *Chem. Lett.* (1994), 901–904.
50. Matsumoto H., Ori A., Inokuchi F., Shinkai S., *Chem. Lett.* (1996), 301–302.
51. Rudkevich D. M. *et al, J. Chem. Soc., Perkin Trans* 2 (1995), 131–134.
52. Steemers F. J., Verboom W., Reinhoudt D. N., van der Tol E. B., Verhoven J. W., *J. Am. Chem. Soc.* **117** (1995), 9408–9414.
53. Steemers F. J., Meuris H. G., Verboom W., Reinhoudt D. N., *J. Org. Chem.* **62** (1997), 4229–4235.
54. Arnaud-Neu F. *et al., Chem. Eur. J.* **5** (1999), 175–186.
55. Malone J. F. *et al, J. Chem. Soc., Chem. Commun.* (1995), 2151–2153.
56. Casnati A., Ungaro R., Arnaud-Neu F., Schwing-Weill M.-J., Dozol J.-F. unpublished results.
57. Wieser Jeunesse C., Matt D., Yaftian M. R., Burgard M., Harrowfield J. M. *C. R. Acad. Sci. Paris* **t1** (1998), 479–502.
58. Arnaud-Neu F. *et al, J. Chem. Soc., Perkin Trans.* 2 (1996), 1175–1182.
59. Delmau L. H., *Chem. Commun.* (1998), 1627–1628.
60. Barboso S. *et al, J. Chem. Soc., Perkin Trans.* 2 (1999), 719–723.
61. Grote Gansey M. H. B. *et al, J. Chem. Soc., Perkin Trans.* 2 (1998), 2351–2360.
62. Arduini A. *et al, Supramol. Chem.* **1** (1993), 235–246.
63. Dasaradhi L., Stark P. C., Huber V. J., Smith P. H., Jarvinen G. D., Gopalan A. S., *J. Chem. Soc., Perkin Trans.* 2 (1997), 1187–1192.

CHAPTER 5

CALIXARENES AS HOSTS FOR QUATS

ANTONELLA DALLA CORT AND LUIGI MANDOLINI

Università "La Sapienza" and Centro CNR di Studio sui Meccanismi di Reazione, Dipartimento di Chimica, Box 34, Roma 62, Italy.

5.1 Introduction

As pointed out by Gutsche[1], the god father of calixarene chemistry, one of the most intriguing properties of this class of compounds is their ability to behave as molecular baskets. Several methods for making, shaping, and embrodering the calixarene baskets are known. This chapter deals with the problem of filling the baskets with quaternary cations. Following a brief outline of the cation-π interaction[2], that plays a key role in this context, this review will consider in the given order complexation of quats by negatively charged and neutral calixarene receptors.

Because of space limitations, the purpose of this review is not to discuss in detail all published calixarene-organic cation complexes, but rather to illustrate the various ways calixarenes have been used as endo receptors for quats.

5.1.1 The cation-π interaction

The various weak noncovalent binding forces that provide the basis for molecular recognition phenomena in biology have been successfully exploited in the design of artificial receptors for molecular recognition in chemistry. Among these forces, the role of the cation-π interaction was generally unappreciated in the past, but is receiving increasing attention in recent years[2]. Numerous experimental observations have established that quaternary ammonium and iminium ions are effectively encapsulated by cyclophane hosts of appropriate geometry and size through interaction of the positive charge of the cation with the π system of the aromatic rings of the host. The importance of the cation-π interaction for biological recognition of a variety of cationic substrates such as acetylcholine and S-adenosylmethionine, has been enphasized.

The essential element for the binding of quats to the π electrons of aromatic compounds is revealed by the prototypical Me$_4$N$^+$-benzene complex **1**, characterized by an experimental free energy of binding $\Delta G° = 3.5$ kcal/mol ($\Delta H° = 9$ kcal/mol) in the gas phase[3], and a calculated distance near 4 Å between the molecular centres.[4,5] There is a generally good agreement between experimental energies for binding of Me$_4$N$^+$ and other monovalent cations to simple aromatic compounds and calculated values from high level theoretical studies.[4,5]

The strong desolvation penalty suffered by the reaction partners upon complexation explains why simple complexes involving a single aromatic ring have been hardly observed in solution studies. Strong binding of quats in solution can be only seen through multiple interactions with a number of strategically placed aromatic rings (multichelate effect), with possible intervention of other interactions.

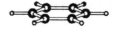

1

Among the various cyclophanes which have been utilized as receptors for organic cations, an important role is played by calixarenes, whose cavity size and shape can be varied and controlled within wide limits. Their utility in the molecular recognition of quats and related cations is the topic of discussion in the remaining part of this chapter.

5.2 Anionic receptors

The importance of binding studies in aqueous media calls for water soluble calixarene receptors. Water solubility is often ensured by a number of polar groups that are negatively charged. Strong binding to quaternary ammonium ions has been reported in many instances, in spite of the substantial desolvation penalty suffered by the cationic guest. Clearly it is difficult to delineate the contribution from the cation-π interaction because other forces contribute to molecular binding, namely, a large hydrophobic effect and classical electrostatic interactions between the solubilizing charged groups and the guest. Nevertheless, in the most recent articles the cation-π interaction has been explicitely acknowledged as an essential component to the overall binding energy.

5.2.1. Deprotonated receptors

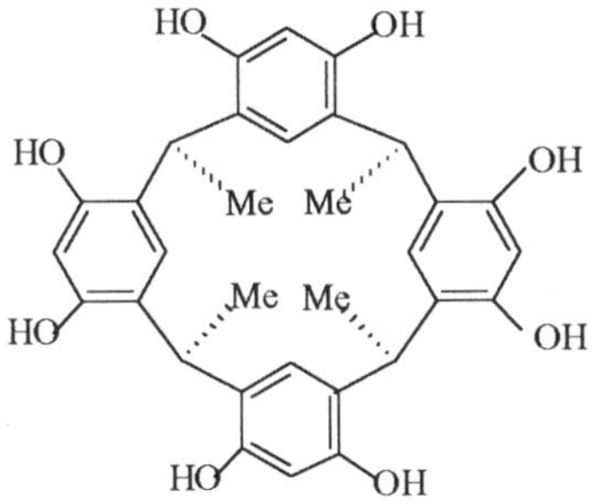

2

The resorcinol-acetaldehyde tetramer **2** is well known to form a water soluble tetraanion 2-H$^+$, rigidly held in a bowl shaped conformation by four intramolecular hydrogen bonds. The tetraanion is water soluble, so that appendend solubilizing groups are not required. In a pioneering investigation of the receptor properties of 2-4H$^+$ toward a series of tetralkylammonium cations in D$_2$O(NaOD) at 25° C Schneider et al.[6] found strong binding energies ΔG° up to 6.4 kcal/mol. A selection of their data is presented in Table 5.1. Interestingly, the strongest association is seen for the biologically relevant cation choline. The large complexation induced ^{1}H NMR shifts (CIS) observed for all the $^+$NCH$_3$ protons and the ordering of CIS values in non-symmetrical cations strongly indicates that in complexes with $^+$NCH$_3$ guest molecules the Me group is located in the cavity of the rather bowllike structure of the host, with the N$^+$ atom lying above the upper rim. Interestingly, the authors attributed the strong CIS values mainly to an effective charge variation in the guest cation upon complexation, with minor contributions from anysotropy effects of the phenolic rings.

Table 5.1. Association constants K (M^{-1}) for receptor **2-4H$^+$** with tetralkylammonium cations in D_2O determined by ^{1}H NMR titration. All CIS values are upfield (-$\Delta\delta_\infty$, ppm).

Cation	K (error)	Proton, CIS
Me$_4$N	29,600 (6,000)	1.84
ET$_4$N	3,400 (600)	CH$_3$, 1.19; CH$_2$, 1.18
Pr$_4$N	30 (5)	1-CH$_2$, 0.42; 2-CH$_2$, 0.42; 3-CH$_3$, 0.42
Bu$_4$N	>2	all H, < 0.01
Me$_3$NCH$_2$CH$_2$OH	50,000 (10,000)	CH$_3$, 2.02; 1-CH$_2$, 1.20; 2-CH$_2$, 0.56
Me$_3$NCH$_2$CH$_2$CH$_3$	40,000 (10,000)	CH$_3$, 2.03; 1-CH$_2$, 1.2; 2-CH$_2$, 0.5; 3-CH$_3$, 0.1
Quinuclidine, Me	5,200 (2,100)	CH$_3$, 2.54; 1-CH$_2$, 1.43; 2-CH$_2$, 0.72; 3-CH, 0.61

A very elegant example of what can be accomplished when an efficient receptor for a given substrate is available, is the study by Inouye et al.[7] The rationale for their

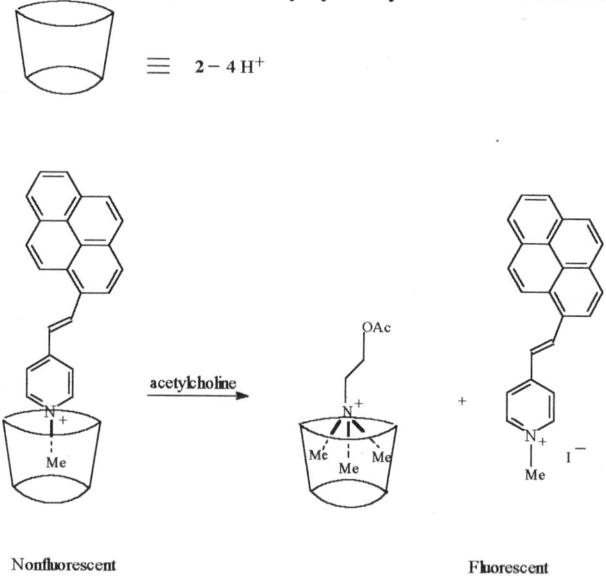

Figure 5.1. A fluorescence-based acetylcholine detection system

artificial-signaling acetylcholine receptor, in which addition of acetylcholine induces a large fluorescene regeneration, is illustrated in Fig. 5.1. The fluorescence of the pyrene-modified N-methylpyridinium dye is strongly quenched by complexation to **2-4H$^+$**, K = 2.3 x 10^5 M^{-1} in 0.01 M KOH/MeOH. Addition of acetylcholine causes fluorescence regeneration, whereas other neurotransmitters do not. The high selectivity observed was attributed to the strong affinity of the receptor to the acetylcholine trimethylammonium head, a structural motif lacking in other neurotransmitters. It has been pointed out[8] that a serious drawback of the above procedure is that both acetylcholine and the fluorescent host are destroyed by the strong alkali required to generate the tetraanion. However, the ingenious strategy has been utilized by the same[7] and other authors[8] to develope modified acetylcholine detection systems operating under neutral conditions (see below).

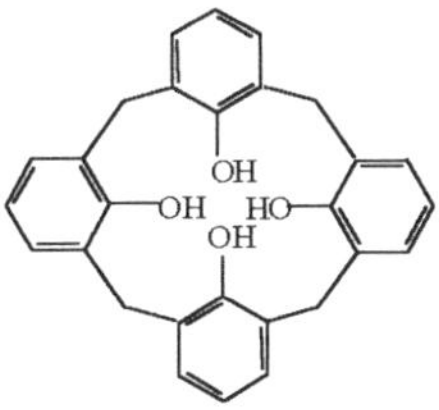

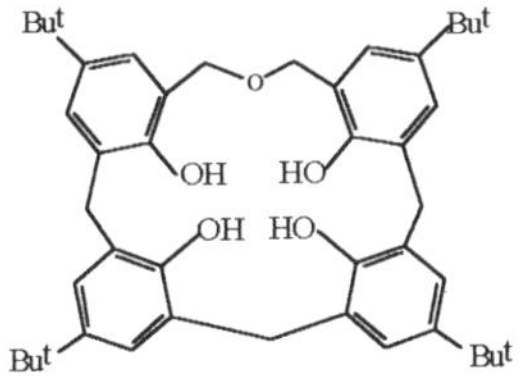

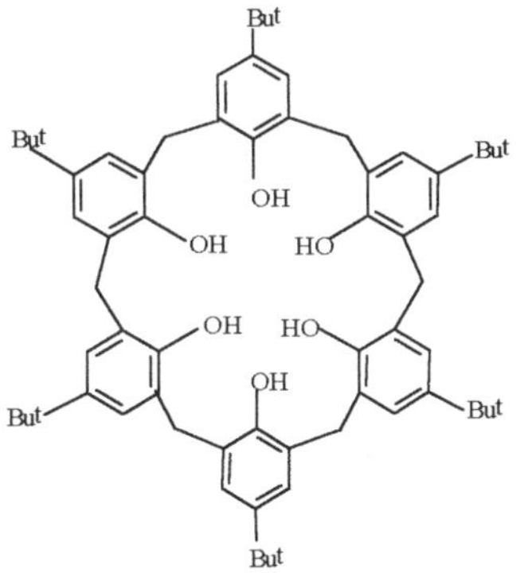

<chemical structures labeled 3, 4, 5>

Strong associations between partially deprotonated calixarenes and tetralkylammonium ions in the aprotic solvents of medium polarity were measured by Harrowfield et al.[9a] (Table 5.2). Qualitative evidence for association was also obtained in the more polar solvent DMSO.[9b] A link between ^{1}H NMR solution data and solid state structures determined by X-ray cristallography was established for some of the investigated salts. Large and negative CIS values of the proton resonances of Me$_4$N$^+$ in the presence of **4-2H$^+$** were taken as evidence for cation inclusion in a way which was believed to resemble that found for the solid salt, where one of the cations is included within a partial cone structure of a "hinged 3-up, 3-down" conformation of the calixarene. Choline and acetylcholine are also strongly included by various calixarene anions, but no evidence for inclusion was obtained with Pr_4N^+ and Bu_4N^+ salts.

Table 5.2. Association constants for calixarenene anions and quaternary ammonium ions.

Cation	Anion	Solvent	K^a, M^{-1}
Choline	4-H$^+$	Acetone[b]	2,500
Choline	3-H$^+$	Acetone[b]	500
Acetylcholine	4-H$^+$	Acetone[b]	850
Acetylcholine	3-H$^+$	Acetone[b]	250
Acetylcholine	5-3H$^+$	Acetonitrile[c]	330,000
Me$_4$N$^+$	5-3H$^+$	Acetonitrile[c]	320,000
Me$_4$N$^+$	5-2H$^+$	Acetonitrile[c]	17,000

a) Errors estimated 20 – 50% ; b) ^{1}H NMR titration, 30° C; c) UV-Vis titration, 25° C.

5.2.2. *Receptors with anionic pendants*

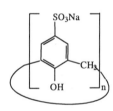

6, n = 4
7, n = 6
8, n = 8

The use of water soluble *p*-sulfonatocalix[n]arenes as complexing agents for quats was pioneered by Shinkai and his group in a series of important papers published in the late eighties and early nineties. The matter has been extensively reviewed by Shinkai himself[10], and only a brief reference is made here.

The principle that calixarenes are capable of molecular recognition on the basis of the ring size was clearly established by the finding that **6** and **7** form 1:1 host-guest complexes with trimethylanilinium chloride, but **8** forms a 1:2 complex. Another important principle established by these studies is that two binding modes (Fig 5.1) are possible with non-simmetrically substituted quats, such as the trimethylanilinium cation. In the complex formed with **6**, it is the phenyl moiety of the trimethylanilinium guest that is located in the basket (structure A) both in the solid state and in D$_2$O solution at pD 0.4. However, at pD 7.3 there is a well balanced equilibrium between the two forms. This was attributed to the change of electron density of the benzene π-system induced by dissociation of one of the OH groups in the neutral solution. Thus, the importance of the cation-π interaction in the recognition of quats by calixarene hosts in aqueos solution was clearly recognized in these early studies.

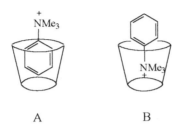

A   B

Figure 5.2. Isomeric structures of a calixarene-trimethylanilinium complex.

In a more recent work,[8] the strong affinity of the octaanion 7-2H$^+$ toward the pyrene-modified N-methylpyridinium dye was exploited for the development of a successful acetylcholine detection system active even in aqueous neutral solution. Other authors have used *p*-sulfonatocalix[n]arenes as receptors for quats in aqueos media. Gokel et al.[11] have investigated the complexation of trialkyl(ferrocenylmethyl)ammonium derivatives by 7-2H$^+$ using electrochemical and ^{1}H NMR spectroscopic techniques. Lehn et al.[12] reported that **6** and **7** complex very strongly the neurotransmitter acetylcholine and other quats in D$_2$O at pD 7.3 with binding constants in the range of 1.6 x·10^3 to 4 x 10^5 M^{-1}. Interestingly, the high affinities measured for choline and acetylcholine (K = 5 x 10^4 to 8 x 10^4 M^{-1}) are comparable to those of the biological recognition sites. The X-ray structure of the complex between **6** and acetylcholine shows that the N-terminal is inserted into the cavity of the receptor in its cone conformation. Calixarenes **6 – 8** are efficiently extracted from water into a chloroform solution of capriquat (trioctylmethylammonium chloride).[13] Interestingly, Mn^{2+} ion is quantitatively extracted into chloroform from an aqueous solution of **6** at pH > 8. By this extraction system, Mn^{2+} was selectively separated from a mixture of a large number of di-, tri-, and tetravalent metal ions.

The receptor properties of resorcin[4]arene with four all-cis CH$_2$CH$_2$SO$_3$Na pendant groups toward a series of ammonium cations in D$_2$O have been described by Aoyama et al.[14] Binding constants (K, M^{-1}) were found to decrease in the order acetylcholine (360), choline (240), Me$_4$N$^+$ (160), Me$_3$NH$^+$ (30), Me$_2$NH$_2^+$ (3), with no evidence for association with both MeNH$_3^+$ and NH$_4^+$ cations. This finding was taken as evidence that, in addition to the hydrophobic effect, it is the CH-π interaction[15], not the cation-π interaction, that makes a substantial contribution to the binding. The authors of this chapter wish to stress that a clean-cut distinction between cation-π and CH-π interactions is clearly applicable when dealing with simple cations such as alkali metal ions, or neutral molecules endowed with CH groups, but with alkylammonium ions the distinction appears to be more semantic than real. On the other hand, solvation provides a satisfactory explanation for the stability decrease on increasing n in the series $Me_{4-n}NH_n^+$, where one finds increasing solvation because of successively more hydrogen-bonding sites.

Carboxylate pendant groups either in place of, or in conjuction with, sulfonate groups have been used by Ungaro and co-workers[16] to achieve water solubility. Receptors 9 and 10, both fixed in the cone conformation, bind $C_6H_5NMe_3^+$ and $C_6H_5CH_2NMe_3^+$ cations t neutral pH with the binding constants listed in Table 5.3.

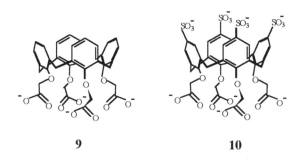

9   **10**

Table 5.3. Binding constants for receptors 9 and 10 in D_2O (pD = 7.3) with quats determined by ^{1}H NMR titration at 25° C.

Host	Guest	K (M^{-1})	CIS (-Δδ$_\infty$, ppm)
9	$C_6H_5NMe_3^+$	160	2.87^b
9	$C_6H_5CH_2NMe_3^+$	50	2.34^b
10	$C_6H_5NMe_3^+$	2000	4.25^c
10	$C_6H_5CH_2NMe_3^+$	2000	1.75^b

a) Errors estimated ± 25%; b) NMe_3^+; c) H_{para}.

^{1}H NMR evidence was obtained that host 9 binds specifically the trimethylammonium head of both guests, as schematically shown by structure B in Fig. 5.2. However, host 10 recognizes only the aromatic ring of $C_6H_5NMe_3^+$ (Fig. 5.2, structure A), whereas binds to $C_6H_5CH_2NMe_3^+$ unselectively at both ends. Table 5.3 shows that complexation is strongly favoured by the four sulfonate groups. Calorimetric measurements, while providing K values in very good agreement with those determined by ^{1}H NMR titrations, showed that the higher stability of complexes with **10** is a result of more favourable $\Delta H°$ contributions.

Another calixarene receptor whose water solubility is ensured by a combination of sulfonate and carboxylate groups is shown in **11** as its complex with $C_6H_5NMe_3^+$.[17]

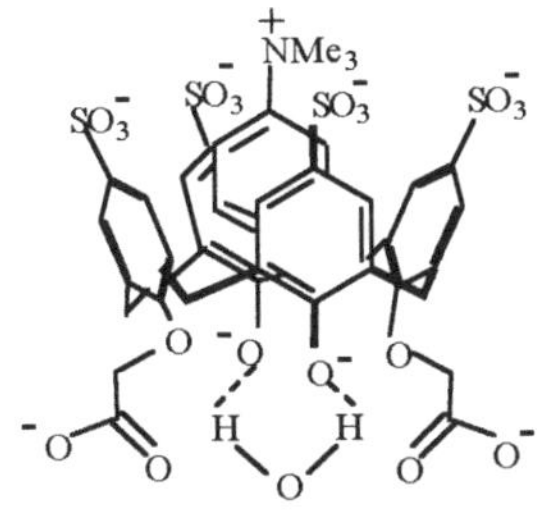

11

The stability of this complex was determined by ^{1}H NMR titration in D_2O (pD = 7.3) at 25° C, K = 2,400 M^{-1}. Compelling evidence that the guest is included in the calixarene cavity as shown in **11** was provided by the large and negative CIS values of the aromatic protons of the guest, whereas a very small upfield shift was experienced by the NMe_3 protons. Analysis of log K values and thermodynamic parameters for the protonation of the phenolate groups obtained by direct calorimetry led to the conclusion that rigidification of the calix conformation is attained through coordination of a water molecule, bridging the two phenolate oxygens.

5.3. Neutral receptors

Considerable importance has been attached to complexation studies involving neutral receptors in organic solvents, where cation-anion interactions and hydrophobic effects are eliminated, and the sole driving force for association is provided by the cation-π interaction. The earliest examples of receptors of this kind are provided by Dougherty's etheno-anthracene-based cyclophane[18] and Collet's cryptophanes[19]. The use of calixarenes as neutral receptors for quats was initiated by Shinkai's group in 1993[20], and the excellent receptor properties of a tetrahomodioxacalix[4]arene derivative were described by De Iasi and Masci[21] in the same year. A number of studies in the field have been published ever since. These studies illustrate the receptor properties of various neutral calix[n]arenes with n = 4 – 6, resorcinarenes, homooxacalixarenes, double calix[4]arenes,as well as the use of calixarenes as building blocks to construct heteroditopic receptors for bifunctional guests.

5.3.1. Calix[n]arenes

Several feature of cation-π interactions in calixarene chemistry were disclosed by Shinkai's group earliest investigation[20]. The ability of a number of calix[n]arenes

12, n = 4
13, n = 6
14, n = 8

15

16

(**12** – **16**) to host quats was qualitatively estimated by ^{1}H NMR spectroscopy, in an attempt to answer the question as to whether the "hole-size relationship" is operating also in cation-π interactions. When octyltrimethylammonium bromide was the guest (CDCl$_3$, 30° C) significant upfield shifts were observed only for the NCH$_3$ and NCH$_2$ protons. Their magnitude varied in the order **15** ~ **13** > **14** and was almost negligible with **12**, **16**, and the p-tBu derivative of **13**, showing that the cavity of a calix[6]arene is best suited to host the charged trimethylammonium head of the guest, and that t-Bu groups on the upper rim interfere with guest inclusion. Additional results were obtained in CDCl$_3$/CD$_3$CN (10:1, v/v) at 30° C with N-methylpyridinium iodide as a guest. Here the largest upfield shifts for the pyridine protons were observed with **12**, whereas the largest upfield shift for the NCH$_3$ protons were observed with **13**. Stability constants (M^{-1}) for receptors **12**, **13**, and **14** are 5.7, 9.8, and 8.3, in the given order[22]. The fraction of cone conformation of the conformationally mobile **12** changed from the natural abundance of 31% to 67% upon addition of 1 mol. equiv. of N-methylpyridinium iodide at –50° C. The conclusion that it is the cone conformation that is stabilized by interaction with the guest was fully confirmed by the finding that only the cone conformation of the conformationally immobile **16**, among four conformational isomers, induces significant upfield shifts of the guest's protons. Complexes formed by trimethylammonium and N-methylpyridinium cations with **12 – 14** and *cone*-**16** are easily detected by positive secondary ion mass spectrometry (SIMS), but the selectivity observed therein and that observed in solution studies are apparently unrelated[23].

In order to reduce the inherent flexibility of calix[6]arene and construct a conformationally-defined host for inclusion of trimethylammonium ions, the basket-like calix[6]arene **17**, C_3-symmetrically capped at the upper rim, was sinthesized[24].

17   **18**

It was found to be an efficient receptor for $C_6H_6NMe_3I$ at low temperatures, K = 65 M^{-1} at 0° C in CD_2Cl_2 (estimated from the van't Hoff plot reported therein). Under the same conditions, the non-bridged reference compound **18** binds to the guest 5.4 times less strongly. The cage structure of **17** causes the complexation-decomplexation process to be slow on the NMR time scale, so that separate signals are seen for the complexed and free guest. In contrast, the more usual behaviour of fast equilibration on the NMR time scale is displayed by **18**. The polyether-bridged calix[6]arene **19** is the result of an alternative strategy aimed at reducing the conformational flexibility via selective bridging followed by alkylation with bulky groups[25]. Compound **19** assumes a cone conformation both in the solid state and in solution, and complexes tetramethylammonium acetate in $CDCl_3$ at 28° C with a remarkably high binding constant, K = 750 M^{-1}. Slow exchange of the ligand causes the appearance of separate signals of the complexed and free host. As to the structure of the host-guest complex, no indication is provided by the signal of the complexed cation because it is hidden by the signals of the host. However, the large shifts suffered upon complexation by the methylenes of the bridge and by the methylenes α to carbonyl strongly suggest that the guest is located in the polar region at the lower rim, and not in the π-base cavity.

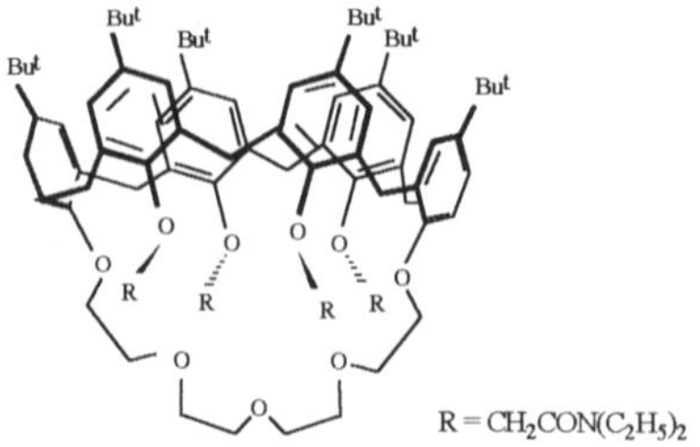

19

R = $CH_2CON(C_2H_5)_2$

Calix[6]arene **20** was synthesized by de Mendoza's group[26] as a monotopic receptor model for a designed ditopic receptor that mimics the phosphocholine binding site of the McPC603 antibody (see below). The high binding affinity of **20** to acetylcholine chloride (K = 138 ± 27 M^{-1} in $CDCl_3$ at 25° C) is surprising, for structural elements deputed to preorganize the macrocycle in the cone conformation are absent.

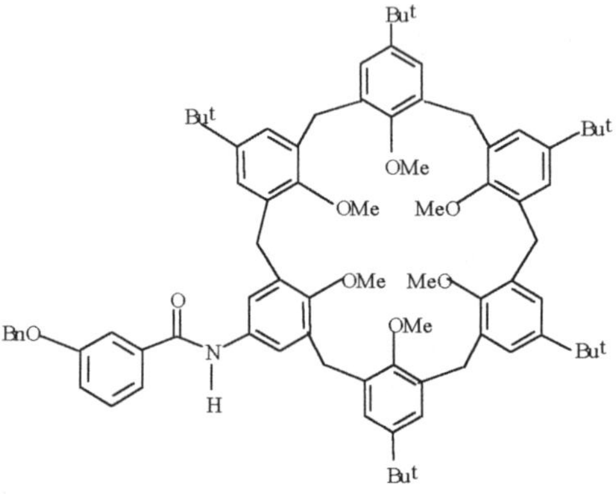

20

The problem of increasing the complexing ability of *cone*-calix[4]arenes toward quats can be tackled in different ways. The strategy followed by Pochini et al.[27] is based on the idea that tetralkoxycalix[4]arenes, blocked in the *cone* conformation

21   **22**

when the alkyl substituents are bulkier than ethyl, are still conformationally mobile, and this residual mobility has an adverse effect on complexation. Their calix[4]arene biscrown **21** provides an extremely rigidified cavity in which the Me_4N^+ cation is efficiently encapsulated. As noted in other instances[21], binding constants (K, M^{-1}, $CDCl_3$) are strongly affected by the counteranion in the order tosylate (33 ± 10), chloride (80 ± 25), acetate (247 ± 3). This is not surprising in view of the extensive ionic association that is likely to occur in solvents of low dielectric constant. The rigid *cone* calix[4]arene π-donor **21** has been successfully exploited by the same group as a building block for the construction of efficient double-calix[4]arene receptors (see below). Shinkai et al.[28] have extended the calix[4]arene system by means of vinyl groups at the upper rim. Enhanced cation-π interactions are shown by the divinyl derivative **22** compared with the parent calix[4]arene **16**. The low binding constant K = 4.2 M^{-1} for complexation of N-methylpyridinium iodide with host **16** rises to 18.4 M^{-1} with **22** ($CDCl_3/CD_3CN$ 4:1, v/v, 25° C).

An insight into the cation-binding properties of the less common calix[5]arenes is provided by an investigation[29] of the complexation of a large number of ammonium, imminium, and phosphonium cations with a series of calix[5]arenes comprising calixcrown **23**, which is known to assume a cone-like conformation in solution, the higher homologue **24**, the de-*tert*-butylated analogue **25**, and the conformationally mobile parent compound **26**. Some of the structure effects involved in cation-binding are disclosed by the selected binding data in Table 5.4. Whereas the length of the polyether bridge has little influence on the binding properties, the adverse effect on complexation of the *tert*-butyl group at the upper rim[20] is fully confirmed and assessed for a number of guests. The conformationally mobile host **26**, unlike hosts **23** and **24**, does not yield complexes of definite stability with most of the investigated hosts, but binds strongly to N-methylquinuclidinium ion, a phenomenon already noticed with a flexible cyclophane receptor[30]. CPK models show that **26**, but not the less flexible **23** and **24**, can adopt a snug-fitting cone conformation ready for close contacts between the five aromatic rings and the globe-shaped guest. In line with previous findings[21,25], the strength of host to guest

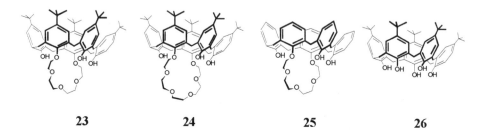

23   **24**   **25**   **26**

Table 5.4. . Stability constants (K, M^{-1}) for the complexes of quaternary ions with hosts **23 – 26** in CDCl$_3$ at 30° C determined by ^{1}H NMR titration.

Cation[a]	23	24	25	26
acetylcholine	47[b,c]	52	210	<5
trimethylanilinium	71	39	210	<5
N-methylquinuclidinium	21	25	81	68
tetramethylphosphonium[d]	87	-[e]	130	-[e]
N-methylquinolinium	24[f]	19	200	<5

a) counterion iodide unless otherwise stated; b) with counterion chloride K = 22 M^{-1}; c) in (CDCl$_2$)$_2$ K = 310 M^{-1}; d) counterion bromide; e) not determined; f) in (CDCl$_2$)$_2$ K = 930 M^{-1}.

binding is strongly dependent on the counterion (footnote b to Table 5.4). The CIS values of the NCH$_3$ protons, -2.33 and -1.52 for the iodide and chloride salt, respectively, indicate that the guest is still bound to its counterion after complexation by the host. The idea that in low polarity solvents the adducts are actually ternary complexes is fully confirmed by a recent study[31a]. The upfield shifts of the acetyl protons of acetylcholine induced by complexation with **25** are affected by the counterion in the same order – tosylate< chloride< iodide< picrate – as the corresponding binding constants. This would indicate that loosening the ion pair the cation is allowed to penetrate more deeply into the host cavity and, consequently, to bind more strongly to the cavity walls. Because picrate ion is a bulky anion with a broadly distributed charge, picrate salts appear to be best suited to afford high binding constants[21]. Tetramethylammonium picrate binds to **23** (CDCl$_3$, 30° C) with a remarkably high affinity, K = 1700 M^{-1}, to be compared with the low binding constant of the tosylate salt, K = 30 M^{-1}, under the same conditions[31a]. Some information of the importance of solvent polarity is provided by the titration data in Fig. 5.3[29], showing that host **23** binds to acetylcholine iodide much more weakly in acetone than in chloroform. There is no evidence for binding in dimethylsulfoxide. The empyrical solvent polarity E$_T$ (30) (kcal mol^{-1}) is 39.1 for chloroform, 42.2 for acetone, and 45.1 for dimethylsulfoxide. It appears therefore that the cation-π interaction cannot overcome the more severe desolvation penalty suffered by host and guest partners in the more polar solvents. The large increase in complex stability on going from chloroform to 1,1,2,2,-tetrachloroethane (footnotes c and f to Table 5.4) was attributed to a lower desolvation penalty suffered by the host in the bulkier 1,1,2,2,-tetrachloroethane, which cannot fit into the calixarene cavity.

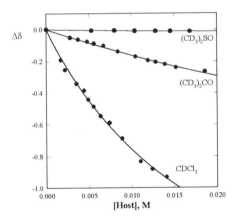

Figure 5.3. ^{1}H NMR titrations (NCH$_3$ protons) of acetylcholine iodide with host **23** in various solvents.

5.3.2. Resorcin[4]arenes

Very few examples of the use of resorcinarenes as neutral receptors for quats have been reported. Since the interaction of **2** with the pyrene-modified N-methylpyridinium cation (see Fig. 5.1) is too weak in neutral protic media, Inouye et al.[7] overcame this problem by preparing compound **27**, soluble in neutral EtOH, in which the fluorescent probe is covalently linked to a neutral resorcinarene receptor. The intramolecularly enforced inclusion of the N-methylpyridinium moiety into the cavity causes fluorescence quenching. Addition of acetylcholine produces a strong emission due to release of the pyridinium moiety from the cavity into the bulk solvent.

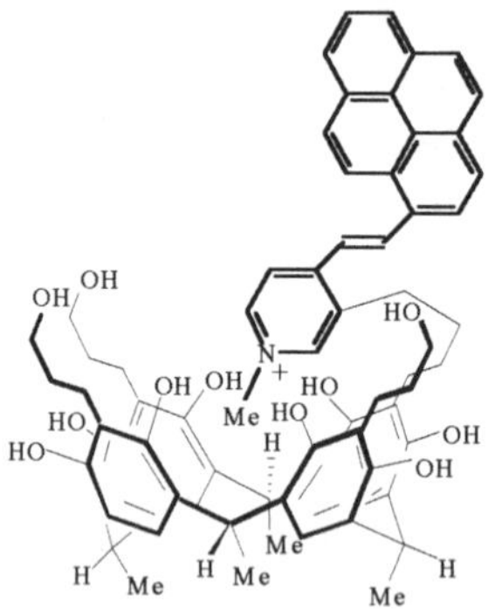

27

Qualitative ^{1}H NMR evidence for inclusion of the trimethyl group of the enantiomerically pure $C_6H_5CH(CH_3)NMe_3^+$ I^- (R-isomer) into the cavity of the chiral resorcinarene receptor **28** in CD_3OD has been produced by Konishi et al.[32]

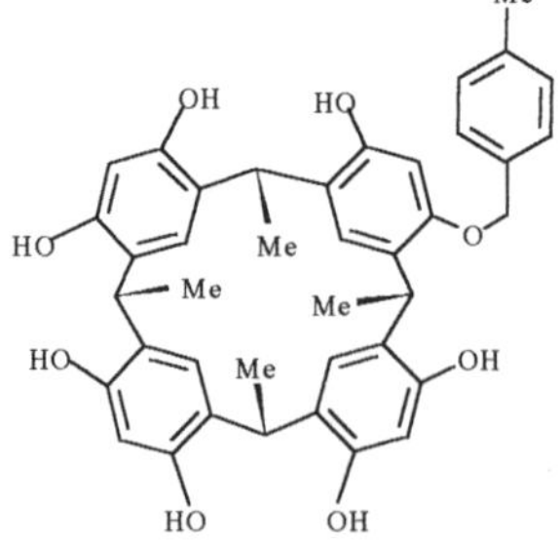

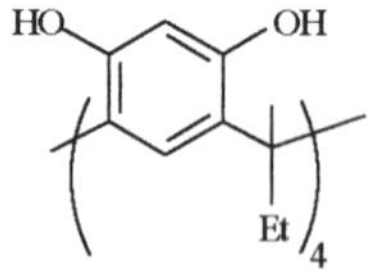

28 **29**

Some of the eight singlets of the aromatic protons of the resorcinol units were split into doublets in the presence of the guest, thus indicating the formation of diastereomeric complexes.

The X-ray crystal structure of a complex between the neutral resorcinol tetramer **29** and acetylcholine chloride has been reported recently[33], (Fig. 5.4). The most interesting structural feature is that two out of the three methyl groups of trimethylammonium head of the choline moiety make close contacts with the four rings of the bowl-shaped cavitand, with C·π–centroid distances in the range of 3.36-3.87 Å. ^{1}H NMR evidence for strong association between acetylcholine and host **29** in CD_3OD solution has also been obtained.

Figure 5.4. Top and side views of the structure of the 1:1 complex between acetylcholine chloride and host **29** · H_2O. (Obtained from the Cambridge Crystallographic Data Centre).

5.3.3. Homooxacalix[n]arenes

The basic idea underlying research into the synthesis of homooxacalixarenes and their use as neutral receptors for quats is that several features of the cyclophanic structure of calix[4]arenes are kept, but larger and tunable cavities are obtained by spacing the aromatic units with groups larger than methylene in one or more sites[34].

The doubly bridged tetrahomodioxacalix[4]arene **30** showed complexation properties towards quats in chloroform, that were unprecedent in the chemistry of calixarenes[21]. The binding constant K (^{1}H NMR titration, 25° C) with N-methylpyridinium iodide is 50 M^{-1}, and with N,N-dimethylpyrrolidinium iodide is 70 M^{-1}. Tetramethylammonium iodide is insoluble in chloroform, but with the soluble acetate salt K = 260^{-1}. The corresponding value for the picrate salt is as large as 1400 M^{-1}. Interestingly receptor **30** binds to tetramethylammonium bromide in EtOH solution with a binding constant of 280 M^{-1} obtained by UV titration.

A large number of quaternary ammonium iodides are complexed by the parent calixarenes analogues **31**– **33** in chloroform solution.[35] As shown by the data listed in Table 5.5., *p-tert*-butylhexaomotrioxacalix[3]arene **31** is rather efficient, but relatively unselective receptor toward a large variety of mono-, di-, and trimethylammonium cations. Receptors **32** and **33** bind to the same cations much less strongly but, interestingly, with larger CIS values almost in all cases. This indicates the lack of a simple relationship between magnitude of the CIS values and binding strength.

Table 5.5. Binding constants K (M^{-1}) and CIS values (-$\Delta\delta_\infty$, ppm) in CDCl$_3$ at 30° C for receptor **31** with quaternary ammonium iodides determined by ^{1}H NMR titrations.

64	90	62	60	62	41	38

Compounds **34-39** constitute a set of cyclophanes in which the size of the potential cavity regularly increases with the number of CH$_2$OCH$_2$ spacers.[36] In spite of the lack of preorganization, as shown by the finding that the cone conformations were disfavoured in general, fairly strong complexes are formed by calixarene-like receptors **35-39** with tetramethylammonium picrate and N-methylpyridinium iodide in (CDCl$_2$)$_2$ solution (Table 5.6). Under the same conditions, no evidence of association was shown by the reference calix[4]arene **34**.

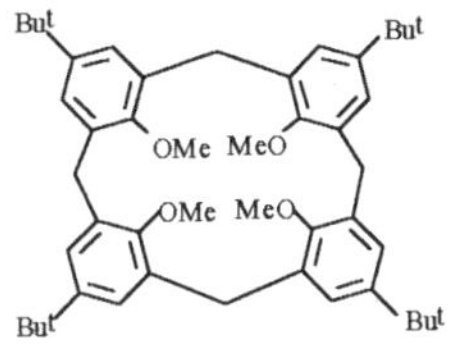

34 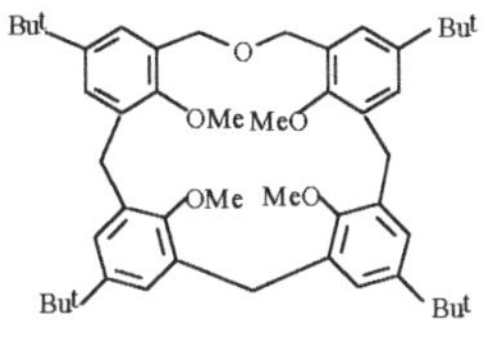   **35** 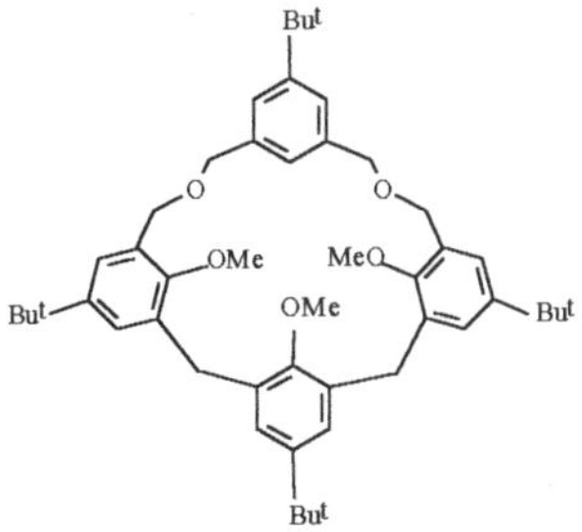   **36**

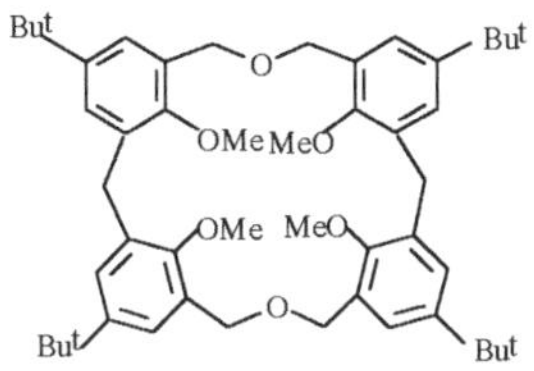

37 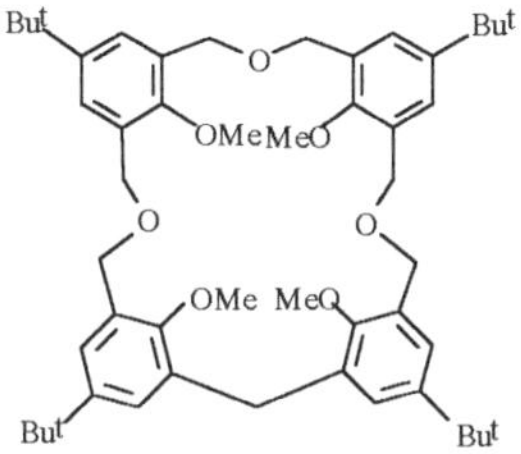   **38** 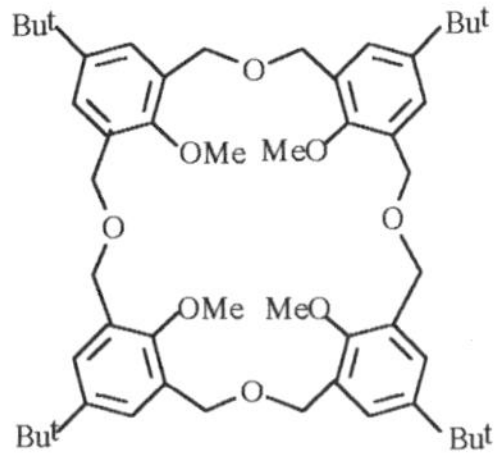   **39**

Table 5.6. Binding constants K(M^{-1}) and CIS values (-Δδ$_∞$, ppm) for receptors **35-39** in (CDCl$_2$)$_2$ at 30° C with tetramethylammonium picrate and N-methylpyridinium iodide determined by ^{1}H NMR.

Receptor	Tetramethylammonium picrate		N-methylpyridinium iodide	
	K	CIS	K	CIS[a]
35	24	0.8	40	1.2, 1.0, 0.5, 0.4
36	270	1.4	120	1.0, 1.1, 0.7, 0.6
37	610	1.6	77	1.0, 1.0, 0.7, 0.6
38	450	1.8	190	1.4, 1.9, 1.4, 1.3
39	470	1.7	190	1.2, 1.6, 1.2, 1.1

a) Signals of Me, α, β, and γ protons in the given order.

5.3.4. Double calix[4]arenes

As is often the case for many facets of calixarene chemistry, the use of double calixarenes as receptors for quats has been pioneered by the group of Shinkai. The bivalve-like receptor **40**[37] in which two *cone*-calix[4]arene subunits are connected at the upper rim by a methylene group, complexes N-methylpyridinium iodide with a binding constant of 480 M^{-1} in CDCl$_3$-CD$_3$CN 5:1, v/v, 24° C. The high degree of cooperativity between the two subunits is indicated by the low binding constant of

6.9 M⁻¹ obtained for the reference compound **16** under the same conditions. Similar results have been reported[38] for the analogous receptor **41** in which the two subunits are connected by means of a *trans* double bond. A high degree of cooperativity is also shown by the bridgeless double calix[4]arene receptor **42**, synthesized by Neri and coworkers.[39] Receptor **42** complexes N-methylpyridinium iodide in $CDCl_3$-CD_3CN 5:1, v/v, 22 °C, with a binding constant of 153 M⁻¹, to be compared with the low value of 6 M⁻¹ found for the reference compound **43**.

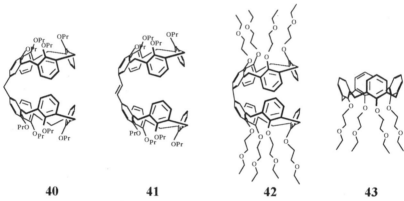

40   **41**   **42**   **43**

The remarkable binding properties of the doubly bridged calix[4]arene **21** have been exploited in the construction of the double calix[4]arene receptors **44 – 46**[40].

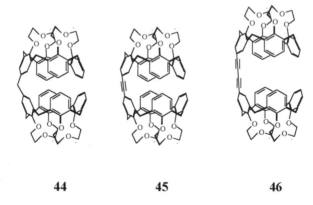

44   **45**   **46**

The equilibrium constants for binding to quaternary pyridinium cations are shown in Table 5.7. The length of, and orientation imposed by the connecting bridge has an important influence on the efficiency of complexation. The best solution is apparently offered by a single triple bond spacer, as shown by the fact that **45** is the best receptor in the lot. On the basis of chemical shift changes upon complexation it has been suggested that the N-methyl-γ-picolinium cation is complexed by **44** with "equatorial" orientation, but with "axial" orientation by **45** and **46**.

Table 5.7. Binding constants K (M^{-1}) for receptors **21**, **44**, **45**, and **46** with quaternary pyridinium iodides in $CDCl_3$-CD_3CN, 9:1, v/v at 27° C determined by ^1H NMR titration.

Guest	21	44	45	46
N-Me-pyridinium	28	200	1600	310
N-Me-α-picolinium	74	100	570	260
N-Me-γ-picolinium	27	120	450	100

The term "sesquicalix[4]arene" is probably more appropriate than double calix[4]arene for the annelated calixarene **47** synthesized by Böhmer et al.[41] Compound **47** turned out to be a remarkably efficient receptor for representative quaternary picrate salts[31b], with a binding constant ($CDCl_3$, 30° C) of 88 M^{-1} with acetylcholine and one of 500 M^{-1} with tetramethylammonium cation. With the N-methylpyridinium cation a value as high as 2200 M^{-1} was obtained. It appears that the flat N-methylpyridinium guest is best suited to fit into the narrow cavity revealed by the computer calculated structure of **47**.

5.3.5. Calixarene-based heteroditopic receptors.

Not much work has been done in this area. However, the two pertinent reports available to date illustrate remarkably well the potential of calixarenes as subunits for the construction of tailor-made supramolecular receptors for highly structured guest molecules. With the idea of obtaining a non-peptidic receptor that mimics the phosphocholine binding site of the McPC603 antibody, de Mendoza et al.[26] have synthesized the ditopic receptor **48**, in which chemical complementarity with the phosphate and trimethylammonium groups of dioctanoyl-L-α-phosphatidylcholine

(DOPC) is provided by the bicyclic guanidinium moiety and calix[6]arene subunit, respectively. Ample NMR evidence was obtained that upon mixing equimolar amounts of **48** and DOPC in either CD_2Cl_2 or $CDCl_3$ a very strong 1:1 complex is formed, in which the amide proton and one of the guanidine protons are hydrogen bonded to phosphate, and the choline trimethylammonium is included into the calix[6]arene cavity. The very high value of 73000 ± 5000 M^{-1} was estimated for the binding constant in chloroform at 25° C. Lower values were obtained for complexes formed by simple model compounds, namely, ca. 7000 M^{-1} for the binding of a simple guanidinium compound to dihexadecylphosphate, and 138 ± 27 M^{-1} for acetylcholine chloride and **20**. These findings clearly show that both guanidinium and calix[6]arene cooperate in binding DOPC but, as the authors themselves have pointed out, there is still room for improvement in the design of the spacer unit. Interestingly, acetylcholine chloride binds very strongly to **48**, $K = 730 \pm 30$ M^{-1}, which in the reviewers' opinion provides an indication that the strength of cation binding is enhanced by hydrogen bonding to the chloride counterion in what amounts to an ion pair recognition process. A similar, but weaker, effect of the same kind is possibly involved in the binding to **20**, where the only hydrogen bond donor group is the amide hydrogen.

DOPC   **48**

The host in the host-guest complex **49** is a *cone*-calix[4]arene-capped tetraphenylporphyrinatozinc in which two distal phenyl *p*-positions of the calix[4]arene subunit are linked to two distal *meso*-phenyl groups of the porphyrinatozinc subunit.[42] The guest is C-(1-methylpyridinium-4-methylamino)-L-isoleucine, used as the BF_4^- salt for solubility reasons. A neutral guest in which the pyridinium moiety is replaced by phenyl was used as reference compound. In

CH$_2$Cl$_2$ the complexation constant with the N-methylpyridinium guest is 7.9 x 10^3 M^{-1}, whereas with the neutral guest is 1.3 x 10^3 M^{-1}. NOE experiments showed that the former is bound inside the cavity by the cooperative action of the cation-π interaction between the charged group and the calix[4]arene, and the coordination of the amino group to the metal centre. In contrast, the monofunctional neutral guests binds to the Zn(II) from the *exo* direction.

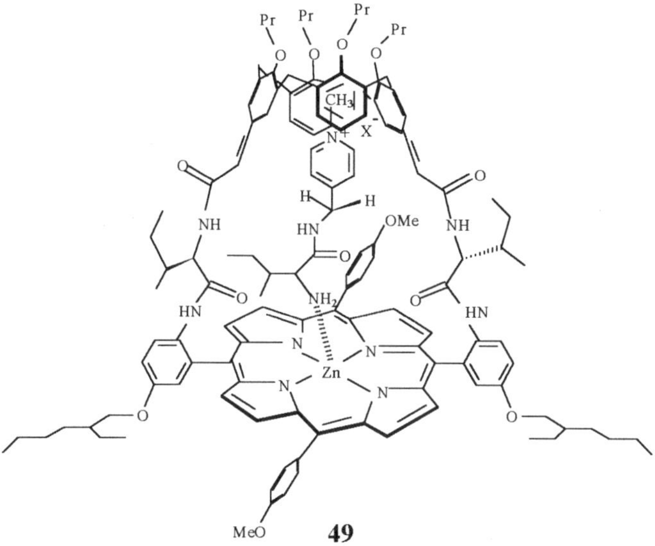

49

5.4. Concluding remarks.

The history of the molecular recognition of quaternary ammonium and iminium ions by calixarene hosts and related compounds is only ten years old. In the early part of this decade the field was dominated by the use of receptors endowed with negatively charged groups at either upper or lower rim of the calixarene backbone. More recent developments have focussed on complexation in nonaqueous media with neutral receptors. Their affinities for quats approach in some cases those of the usually more efficient negatively charged systems. The importance assumed by the calixarenes in the study of complexations driven by cation-π interactions now rivals with that of other cyclophane systems. The very latest improvements in the field, namely, the construction of efficient homo and heteroditopic receptors, clearly indicate that there is much scope for further applications of calixarenes in the molecular recognition of quats. This is particularly so in their potential as prefabricated parts for the construction of polytopic receptors of increasing complexity.

5.5. References

1. a) Gutsche C. D., *Calixarenes* (The Royal Society of Chemistry, Cambridge, 1989); b) Gutsche C. D., *Calixarenes Revisited* (The Royal Society of Chemistry, Cambridge, 1998)
2. Ma J.C., Dougherty D.A., *Chem. Rev.*, **97**, (1997), 1303-1324.
3. Meot-Ner (Mautner) M., Deakyne C. A., *J. Am. Chem. Soc.*, **107**, (1985), 469-474.
4. Duffy E.M., Kowalczyk P.J., Jorgensen W. L., *J. Am. Chem. Soc.*, **115**, (1993), 9271-9275.
5. Pullman A., Berthier G., Savinelli R., *J. Comput. Chem.*, **18**, (1997), 2012-2022.
6. Schneider H.-J., Güttes D., Schneider U., *J. Am. Chem. Soc.*, **110**, (1988), 6449-6454.
7. Inouye M., Hashimoto K., Isagawa K., *J. Am. Chem. Soc.*, **116**, (1994), 5517-5518.
8. Koh K. N., Araki K., Ikeda A., Otsuka H., Shinkai S., *J. Am. Chem. Soc.*, **118**, (1996), 755-758.
9. a) Harrowfield J.M., Richmond W.R., Sobolev A. N., *J. Inclusion Phenom. Mol. Recognit. Chem.*, **19**, (1994), 257-275. b) Harrowfield J. M., Ogden M. I., Richmond W. R., Skelton B. W., White A. H., *J. Chem. Soc. Perkin Trans. 2*, (1993), 2183-2190.
10. Shinkai S., *Tetrahedron*, **49**, (1993), 8933-8968.
11. Zhang L., Macias A., Lu T., Gordon J.I., Gokel G. W., Kaifer A. E., *J. Chem Soc. Chem. Comm.*, (1993), 1017-1019.
12. Lehn J.-M., Meric R., Vigneron J.-P., Cesario M., Guilhem J., Pascard, C., Asfari Z., Vicens J., *Supramol. Chem.*, **5**, (1995), 97-103.
13. Nishida M., Sonoda M., Ishii D., Yoshida I., *Chem. Lett*, (1998), 289-290.
14. Kobayashi K., Asakawa Y., Aoyama Y., *Supramol. Chem.*, **2**, (1993), 133-135.
15. Nishio M., Hirota M., Umezawa Y., *The CH/π Interaction* (Wiley-VCH, New York, 1998).
16. a) Arduini A., Casnati, A., Fabbi P., Minari P., Pochini A., Sicuri A.R., Ungaro R., *Supramolecular Chemistry* (V. Balzani, L. De Cola Eds, Kluwer,

Dordrecht, 1992), 31-70; b) Arena G., Casnati A., Contino A., Lombardo G.G., Sciotto D., Ungaro R., *Chem. Eur. J.*, **5**, (1999), 738-744.
17. Arena G., Casnati A., Mirone L., Sciotto D., Ungaro R., *Tetrahedron Lett.*, **38**, (1997), 1999-2002.
18. Stauffer D.A., Dougherty D.A., *Tetrahedron Lett.*, **29**, (1988), 6039-6042.
19. Collet A., Dutasta J.-P., Lozach B., *Bull. Soc. Chim. Belg.*, **99**, (1990), 617-633.
20. Araki K., Shimizu H., Shinkai S., *Chem. Lett.*, (1993), 205-208.
21. De Iasi G., Masci B., *Tetrahedron Lett.*, **34**, (1993), 6635-6638.
22. Binding affinities are quoted in ref. 20 as log K and the corresponding $\Delta G°$ values (kcal mol^{-1}). Consistency between the two sets of values is only obtained if log K is taken as ln K (*not* log$_{10}$ K). See also footnote 17 in ref. 35. Similar considerations most likely apply to the log K values given in refs. 23 and 38.
23. Inokuchi F., Araki K., Shinkai S., *Chem Lett.*, (1994), 1383-1386.
24. Takeshita M., Nishio S., Shinkai S., *J.Org. Chem.*, **59**, (1994), 4032-4034.
25. Casnati A., Iacopozzi P., Pochini A., Ugozzoli F., Cacciapaglia R., Mandolini L., Ungaro R., *Tetrahedron*, **51**, (1995), 591-598.
26. Magrans J. O., Ortiz A. R., Molins M. A., Labouille P. H. P., Sanchez-Quesada J., Prados P., Pons M., Gago F., de Mendoza J., *Angew. Chem. Int. Ed. Engl.*, **35**, (1996), 1712-1715.
27. Arduini A., McGregor W. M., Paganuzzi D., Pochini A., Secchi A., Ugozzoli F., Ungaro R., *J.Chem. Soc. Perkin Trans. 2*, (1996), 839-846.
28. Lhoták P., Nakamura R., Shinkai S., *Supramol. Chem.*, **8**, (1997), 333-344.
29. Arnecke R., Böhmer V., Cacciapaglia R., Dalla Cort A., Mandolini L., *Tetrahedron*, **53**, (1997), 4901-4908.
30. Cattani A., Dalla Cort A., Mandolini L., *J. Org. Chem.*, **60**, (1995), 8313-8314.
31. a) Arnecke R., Böhmer V., Dalla Cort A., Mandolini L., unpublished results; b) Böhmer V., Frings M., Dalla Cort A., Mandolini L., unpublished results.
32. Konishi H., Tamura T., Ohkubo H., Kobayashi K., Morikawa O., *Chem. Lett.*, (1996), 685-686.
33. Murayama K. and Aoki K., *J. Chem. Soc. Chem. Comm.*, (1997), 119-120.
34. Masci B., Saccheo S., *Tetrahedron*, **49**, (1993), 10739-10748.
35. Masci B., *Tetrahedron*, **51**, (1995), 5459-5464.
36. Masci B., Finelli M., Varrone M., *Chem. Eur. J.*, **4**, (1998), 2018-2030.
37. Araki K., Hisaichi K., Kanai T., Shinkai S., *Chem. Lett.*, (1995), 569-570.
38. Lhoták P., Shinkai S., *Tetrahedron Lett.*, **37**, (1996), 645-648.

39. Neri P., Bottino A., Cunsolo F., Piattelli M., Gavuzzo E., *Angew. Chem. Int. Ed. Engl.*, **37**, (1998), 166-169.
40. Arduini A., Pochini A., Secchi A., unpublished results.
41. Böhmer V., Dörrenbächer R., Frings M., Heydenreich M., de Paoli D., Vogt W., Ferguson G., Thondorf I., *J. Org. Chem.*, **61**, (1996), 549-559.
42. Nakamura, R., Ikeda A., Sarson L. D., Shinkai S., *Supramol. Chem.*, **9**, (1998), 25-29.

CHAPTER 6

CALIXARENE BASED ANION RECEPTORS

PAUL D. BEER AND JAMES B. COOPER

6.1 Anion Co-ordination Chemistry

Anions are ubiquitous in biological systems[1, 2] and are significant agricultural and industrial pollutants. For example, nitrates and phosphates from fertilisers cause eutrophication in rivers[3] while TcO_4^- is a dangerous waste product of the nuclear industry. Anions are also of importance in many illnesses, such as in the misregulation of chloride channels that leads to cystic fibrosis[4]. Despite this the non-covalent binding of anionic guest species was for a long time a relatively unexplored field. In recent years the need to develop a means of sensing and extracting polluting anions has led to an intense interest in anion co-ordination chemistry and calixarenes have played an important role in the development of this area.

The aim of this chapter is to describe the various ways calixarenes have been utilised as anion receptors. Following an initial discussion of the challenges of anion binding and a brief historical overview, the role of calixarenes as anion receptors will be examined in detail. For the purpose of the following discussion the calixarene based anion receptors will be separated into charged and neutral classes.

6.1.1 The Challenge of Anion Binding

The birth of anion co-ordination chemistry came with the synthesis by Simmons and Park[5] of the first artificial anion receptor in 1968, only a year after the discovery of crown ethers[6]. However it is only recently that the field of anion co-ordination has developed to an extent approaching that of cation co-ordination. This may appear quite surprising, as anions are as fundamental as cations in biological processes. For example chloride, like sodium and potassium ions, acts as an osmotic electrolyte, requiring ion channels through the cell membrane[7]. Part of the reason for this difference may be ascribed to the greater challenge presented to the supramolecular chemist by anions than cations. Although the approaches learnt from cation co-ordination chemistry are important there are some fundamental differences between cations and anions that need to be taken into account in the designing of a potential anion receptor.

(a) Charge - This is the most obvious but nonetheless important difference as it affects the choice of functional groups utilised for binding; e.g. an amine may bind cations whereas a protonated amine may co-ordinate anions.

(b) Size - Anions are generally much larger than isoelectronic cations due to the anion's smaller effective nuclear charge (**Table 6.1**). Anions containing more than one atom are also very common. This means that the binding sites for anions need to be significantly larger than for cations.

Table 6.1 Comparison of ionic radii[8] in the solid state for various isoelectronic cation/anion pairs.
[a]six co-ordinate in the MX crystal.

Isoelectronic cation/anion pair	Cation ionic radius[a]/pm	Anion ionic radius[a]/pm
K^+/Cl^-	138	184
Rb^+/Br^-	152	196
Cs^+/I^-	167	220

(c) Shape - Anions, unlike cations which are generally spherical, exhibit a wide variety of geometries. This imparts greater significance on the spatial arrangement of the binding groups within the binding site. (**Figure 6.1**)

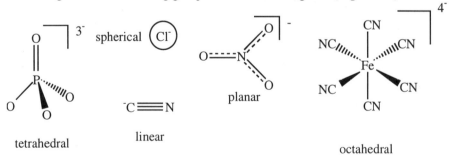

Figure 6.1 **Common shapes of anions.**

(d) pH Dependence - Unlike cations many anions only exist over a limited pH range. For example carboxylate anions exist only above pH 5 or 6.

(e) Solvation - The degree of solvation of anions has a large effect on the strength of interactions, thus the solvent can have a dramatic effect on the strength and selectivity of binding[9].

Nature copes with all of these requirements very successfully in the intricate binding sites of anion selective proteins[10]. Recent X-ray crystallographic studies[11] have revealed the complex arrays of hydrogen bond donor/acceptor groups used to selectively bind anions such as phosphate and sulphate in proteins where the binding site is perfectly configured for the shape and size of substrate. For the supramolecular chemist the task of selectively recognising a target anionic guest species is a difficult one, the above considerations all play a role in the development of anion receptors.

6.1.2 The Binding Site

Several strategies have been developed for the binding of anions[2, 12]. These include:

(a) Hydrogen bonding - Anions interact with polar H-X bonds (X=N, O, S etc.) to form hydrogen bonds. Thus amines, amides, alcohols and thiols can all act as potential anion binding sites and have been widely utilised as such. Hydrogen bonding is the primary source of anion binding in nature[10], in artificial receptors it can be used as the sole binding interaction but it is often employed in combination with other strategies.

(b) Electrostatic attraction - The short range attraction between an anion and a positively charged group is an obvious and common strategy. There are several possible methods of including charge in the binding site: protonated or quaternised amines, inclusion of a charged organic groups e.g. guanidinium or by the presence of a co-bound metal cation.

(c) Lewis Acids - An electron deficient or Lewis acid centre is used to attract the electron rich anion. Commonly used Lewis acid centres include organo-tin, boron, uranyl and transition metal moieties.

6.1.3 A Historical Perspective

The electrostatic attraction of a positively charged group is probably the most obvious means of attracting and binding anions, and consequently it was one of the first to be exploited.

As previously mentioned one of the first molecules to exhibit anion co-ordination was synthesised by Simmons and Park[5] (**1a**) in 1968

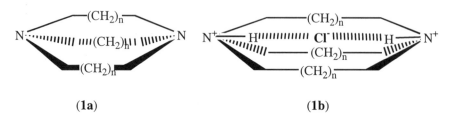

(**1a**) (**1b**)

When the nitrogen atoms are protonated (**1b**) a halide ion (e.g. chloride when n=9) is bound in the cage by a combination of electrostatic attraction and hydrogen bonding. The strength and selectivity of halide binding is determined by the size of the cage.

It was several years later that further molecules were made utilising protonated amines to bind anions. In 1976 Lehn[13] described the encapsulation of spherical halide anions in the protonated form of the macrotricycle (**2**).

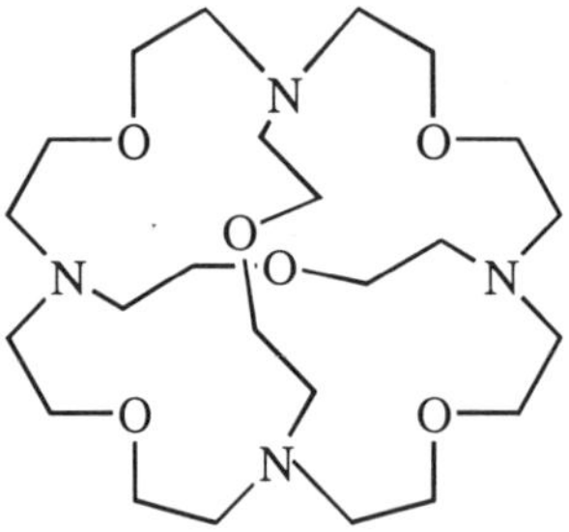

(**2**) 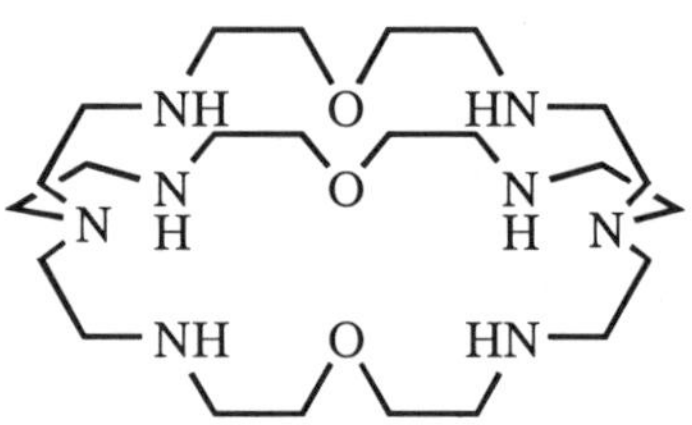 (**3**)

The cryptand (**3**) is another example of a pH dependent polyammonium macrocycle. When hexaprotonated it binds anions, with a high selectivity for azide[14]. This receptor demonstrates another important finding to come from the early work on anion binding, the importance of host-guest shape complementarity. The selectivity of receptor (3) for azide can be attributed to the elliptical shape of its cavity matching the size and shape of the azide anion.

The problem with this kind of polyammonium macrocycle is its pH dependence; there is only a limited pH range over which the macrocycle is fully protonated and the anion is not. In an attempt to overcome this problem the pH independent quaternised macropolycycle (4) was prepared[15]. The cavity size could be varied affecting the selectivity, for example with n=6 this receptor selectively bound bromide. Due to the lack of hydrogen bonding these quaternary nitrogen based macrocycles formed relatively weaker complexes than the protonated polyammonium analogues.

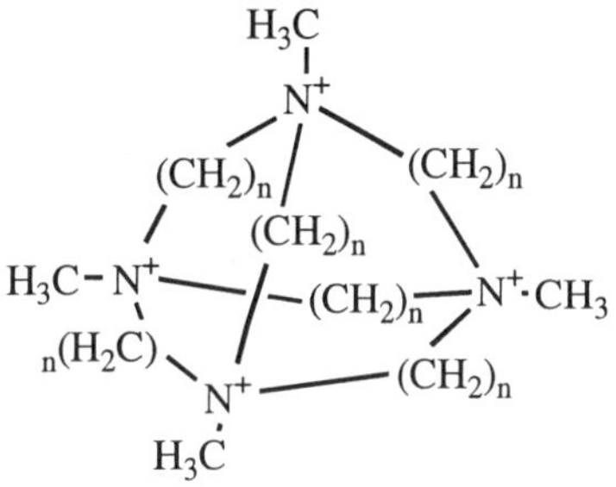

(4)

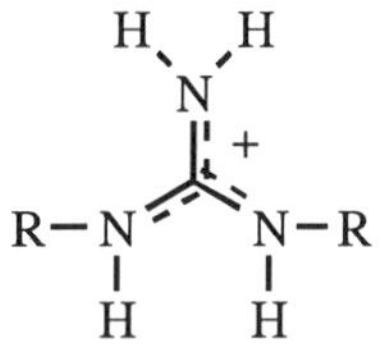

(5)

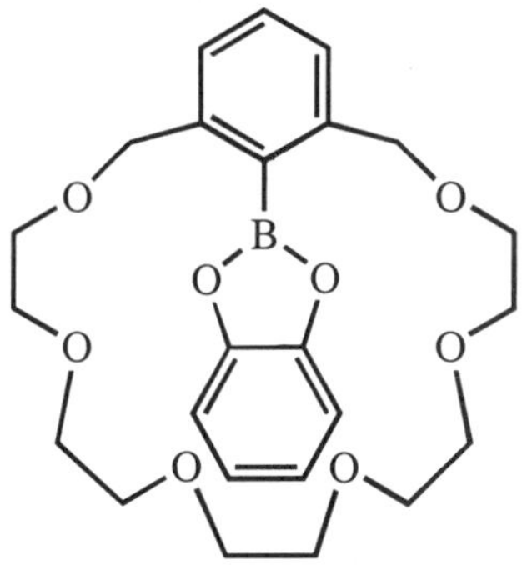

(6)

Another way to overcome the problem of pH dependent receptors was developed using the charged guanadinium group[16] (5). These were once again inspired by nature where guanadinium is often used in enzymatic anion-binding systems in the form of arginine residues[17]. The guanidinium moiety has a pK_a=12-13 depending on its substituents and thus is able to bind anions over a much wider pH range than polyammonium receptors, whilst retaining the hydrogen bonding ability that is lost in the quaternised receptors. Further developments in anion binding have lead to the use of the essentially pH independent amide moiety to form hydrogen bonds to anions, many such examples of calixarene based systems appear later in this chapter. Neutral Lewis acid centres, such as tin[18], silicon[19], germanium[20], mercury[21] and boron as in receptor (**6**) which is capable of ion pair binding[22], have been used to great effect in non-calixarene anion receptors. These groups have yet to be exploited in conjunction with calixarenes, however other Lewis acid centres have been used in calixarene anion receptors and will be described later in this chapter.

6.1.4 Calixarene Based Anion Receptors

Calixarenes provide a unique platform for the supramolecular chemist to build on. Their semi-rigid, three dimensional structure allows the positioning of anion binding sites to be matched with the size and shape of the target anion. The upper and lower rims are easily modified with a wide variety of functional groups, and in addition the bowl of the calixarene and the oxygen atoms on the lower rim can form an important part of the binding site.

The remainder of this chapter will examine the different ways in which calixarenes have been modified as receptors for anions. They will be subdivided under the classification of charged and neutral types.

6.2 Charged Calixarene Anion Receptors

The field of anion chemistry had developed quite extensively by the time calixarenes were first used as anion receptors. In creating charged receptors the majority of strategies have involved the use of pH independent metal centres to carry the positive charge. These can be attached via a linking group or directly to the calixarene to form the binding site. Arguably the first calixarene based anion receptors containing a charged centre were cobaltocenium calixarenes

6.2.1 Cobaltocenium Calix[4]arene Anion Receptors

One of the earliest examples of this type of receptor utilised the positively charged pH independent cobaltocenium moiety attached to the upper rim of calix[4]arene (**7**)[23], more recently an upper rim bridged form (**8**) has been synthesised[24]. These molecules have the added advantage of being redox active resulting in the capability to sense anion binding via electrochemical methodologies.

Receptor (**7**) has been shown by ^{1}H NMR titration techniques to strongly bind halide anions in d_6-DMSO solution and to form a strong 1:1 complex with the dicarboxylate dianion $^-O_2C(CH_2)_4CO_2^-$ (adipate) in d_6-acetone. Significant perturbations of the calix aryl protons are observed upon complexation indicating that binding is occurring within the upper rim of the calixarene cavity. The two positive charges are set up to co-operatively bind one end of the dicarboxylate anion with the aliphatic chain possibly residing in the hydrophobic calixarene cavity.

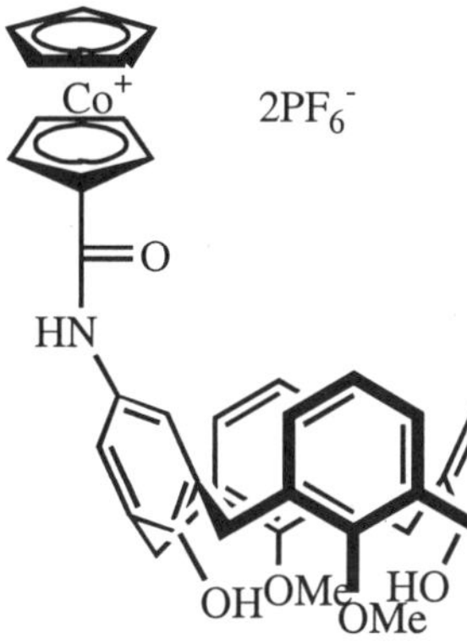

(**7**)

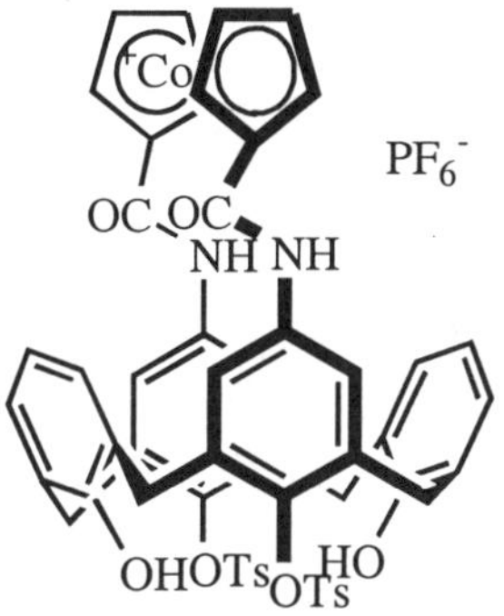

(**8**)

Receptor (**8**) also forms strong complexes with anions in d_6-DMSO solution with a marked selectivity for mono carboxylates over other anions such as halides, nitrate and dihydrogen phosphate. The bridged calixarene amide cobaltocenium framework results in the preorganisation of the hydrogen bond donating amides for anion binding. This can be seen in selectivity for RCO_2^- bidentate anions and in the X-ray crystal structure of the chloride complex of (**8**) (**Figure 2**)

A model cobaltocenium compound in which the calixarene is replaced with non-bridging phenyl groups shows much weaker anion binding e.g. for acetate K=1500 M^{-1} for the model vs K=41500 M^{-1} for receptor (**8**) demonstrating the importance of the calixarene in preorganising the binding site.

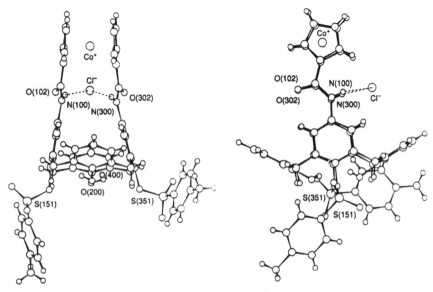

Figure 6.2 Crystal structure of receptor (**8**) chloride complex

Cobaltocenium is redox active and addition of anions causes a significant perturbation of the cobaltocenium/cobaltocene redox couple allowing the receptors to act as sensors for anions. Binding an anion in close proximity to the cobaltocenium moiety disfavours the reduction from Co(III) to Co(II) resulting in a cathodic shift in the reduction potential compared to the free receptor. Large shifts are observed for receptor (**8**) with carboxylate anions, which is consistent with the carboxylate anion selectivity trends determined by ^{1}H NMR anion binding studies. (**Table 6.2**).

Table 6.2 Electrochemical data for (**8**) [a]obtained in acetonitrile solution containing 0.1 M^{-1} nBu$_4$NPF$_6$ as supporting electrolyte. [b] Cathodic shift perturbations of cobaltocenium/cobaltocene redox couple of (**8**) produced by the presence of anions (up to 5 equiv.) added as their tetrabutyl ammonium salts.

Anion	ΔE^b/mv
Cl$^-$	60
MeCO$_2^-$	155
PhCO$_2^-$	140

The overall calixarene structure has an important influence on anion binding, substitution of the lower rim with bulky tosyl groups can affect the spatial arrangement, and thus binding ability of, an upper rim anion recognition site. This is demonstrated by comparing the structural isomers of the cobaltocenium receptors (**9a, 9b**) [25].

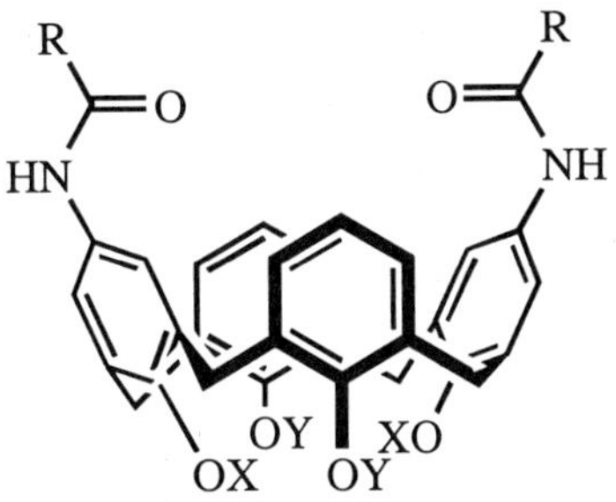

	R	X	Y
(**9a**)	$Co(Cp)_2^+$	Tos	H
(**9b**)	$Co(Cp)_2^+$	H	Tos

Table 6.3 Stability constants for receptors (**9a**) and its structural isomer (**9b**) in d_6-DMSO determined by ^1H NMR titration. [a] Errors estimated to be ≤ 10%. [b] Binding to weak to determine stability constant.

	K/M^{-1} [a]	
Receptor	$H_2PO_4^-$	Cl^-
(**9a**)	3100	[b]
(**9b**)	2500	400

As can be seen from **Table 6.3** the substitution pattern on the lower rim has a dramatic effect on the strength of anion binding. Comparing (**9a**) to its structural isomer (**9b**) there is a large decrease in the strength of phosphate binding and enhancement of chloride binding when the free phenol is para to the amide substituent rather than the tosyl group. Molecular modelling indicates that in (**9a**) the effects of the lower rim bulky tosyl groups result in a receptor conformation where the upper rim charged centres are held in close proximity which in turn favours phosphate binding.

6.2.2 Ruthenium(II) Bipyridyl Calix[4]arene Receptors

Ruthenium(II) bipyridyl, like cobaltocenium, is pH independent and redox active but also has the added advantage of being photoacitve and so it can be used for both optical and electrochemical sensing for anions. The charged ruthenium fragment Ru(bipy)$_3^{2+}$ has been appended to both the upper[25, 26] and lower rims of calixarenes[26, 27] by Beer et al. Ruthenium bipyridyl analogues of receptors (**9a**) and (**9b**) (R=Ru(bipy)$_3^{2+}$) have been synthesised and display similar dependence of anion binding strength on lower rim substitution patterns.

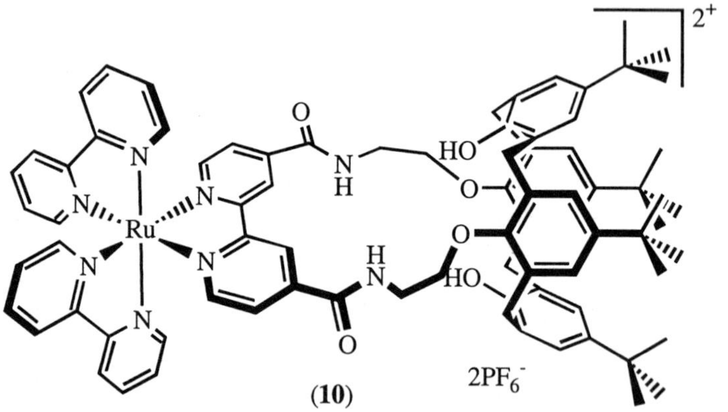

Receptor (**10**) shows selectivity for H$_2$PO$_4^-$ in d_6-DMSO over halides. (**Table 6.4**). Perturbation of the calixarene hydroxyl proton resonances in the ^{1}H NMR spectrum suggests that they are involved in binding and the solid state X-ray crystal structure of the phosphate complex supports this. The calixarene forms an important part of the binding site, along with the amide protons and the electrostatic attraction of the metal.

Table 6.4 Stability constants for receptor (**10**) in d_6-DMSO determined by ^{1}H NMR titration. ^{a} Errors estimated to be ≤ 5%.

Anion	K^a (M^{-1})
Cl$^-$	1600
Br$^-$	360
H$_2$PO$_4^-$	28000

The ruthenium(II) bipyridyl moiety can, like cobaltocenium, act as an electrochemical probe. In a competition experiment[26] with a ten fold excess of Cl$^-$ and HSO$_4^-$ over H$_2$PO$_4^-$ in acetonitrile, the bipyridyl ligand reduction couple was shifted cathodically by an amount approximately the same as that induced by the $H_2PO_4^-$ anion alone, suggesting evidence of selective phosphate electrochemical recognition. Fluorescence studies[27] carried out in DMSO show a marked enhancement in emission intensity of receptor (**10**) in the presence of $H_2PO_4^-$ anions indicating its ability to act as a fluorescent anion sensor as well as an electrochemical one.

The bis calixarene ruthenium(II) bipyridyl receptor (**11a**) when compared to the non-calixarene analogue (**11b**) illustrates the dramatic effect the proximal calixarenes have on the strength of binding of various anions[28].

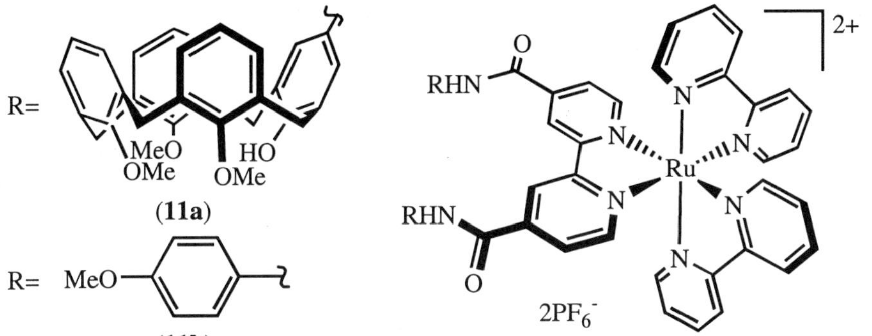

Receptor (**11b**) displays the anion selectivity trend usually found for this type of receptor i.e. binding harder anions in preference to softer charge diffuse anions. The bis calix[4]arene receptor (**11a**) however displays a marked decrease in stability for the hard acetate, and dihydrogen phosphate and an increase in the strength of binding of the more diffuse benzoate and phenyl acetate guest species. (**Table 6.5**)

Table 6.5 Stability constants for receptor (**11a**) and (**11b**) in d_6-DMSO determined by ^1H NMR titration.
[a] Errors estimated to be ≤ 15%.

Anion	K^a (M^{-1})	
	(**11a**)	(**11b**)
Cl$^-$	145	220
$H_2PO_4^-$	630	2000
$MeCO_2^-$	160	1200
$PhCO_2^-$	750	300
$PhCH_2CO_2^-$	650	580

Differences in solvation of the host or the host-guest anion complex due to the presence of the calixarene most likely account for the selectivity change. Once again the presence of the calixarene is important in determining the strength and selectivity of anion binding.

6.2.3 π-Metallated Calixarenes

In previous examples the positively charged organometallic co-ordination metals have been attached to the calixarene via linking groups. An alternative approach has been adopted by Atwood *et al.* who synthesised π-metalated derivatives of calixarenes with the metals actually co-ordinated to the faces of the calixarene bowl. Di- and tetra-substituted derivatives of calix[4]arene with Rh, Ir, and Ru and a tri-Ir complex of calix[5]arene have been synthesised. [29]

The electron withdrawing nature of the cationic organometallic transition metals significantly enhances the acidity of the phenolic protons resulting in the calixarene (**12a**) being doubly deprotonated. The withdrawal of electron density from all four aryl rings in the calixarene along with the overall high charge is enough to cause it to crystallise with one of the BF_4^- counter ions within the calixarene cavity[30].

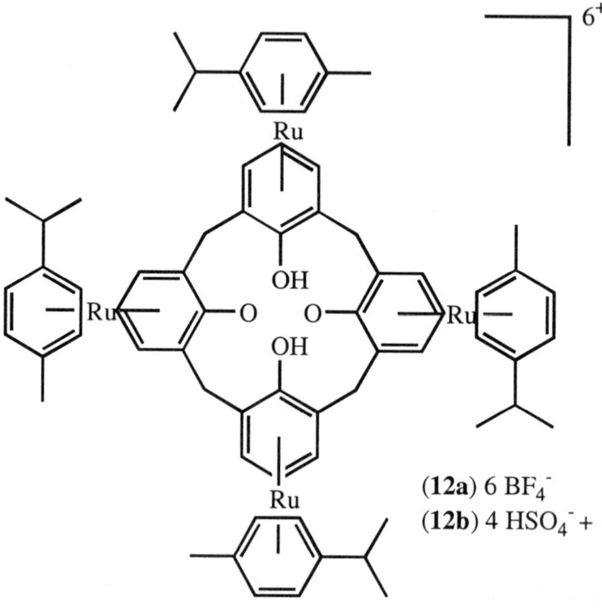

(**12a**) 6 BF_4^-
(**12b**) 4 HSO_4^- + SO_4^{2-}

The tri- and tetrametalated derivatives were all found to include anions in the molecular cavity in the solid state. The tetra-Ru compound (**12b**) was found to bind halide and nitrate anions in water. (**Table 6.6**)

Table 6.6 Stability constants (M^{-1}) for receptor (**12b**) water determined by ^1H NMR titration. [a] Errors estimated 10%.

Anion	K_1
Cl^-	551
Br^-	133
I^-	51
NO_3^-	49

There is a significant decrease in the strength of binding as the ionic radius increases from chloride to bromide. The degree of binding is influenced by the size of the cavity. The larger nitrate ion is less able to fit into the cavity but has a K_2 (109 M^{-1}) which is larger than K_1 possibly due to the planar nitrate bridging two metal centres outside the calixarene bowl.

Analogous receptors have also been synthesised using cyclotriveratrylene (CTV). The di-ruthenium CTV receptor (**13**) has been shown to complex ReO_4^- anions. Binding ReO_4^- is an important development due to the importance of the topologically similar TcO_4^- as a pollutant from the nuclear industry. One possible application of these receptors is in the selective complexation and removal of such pollutants.

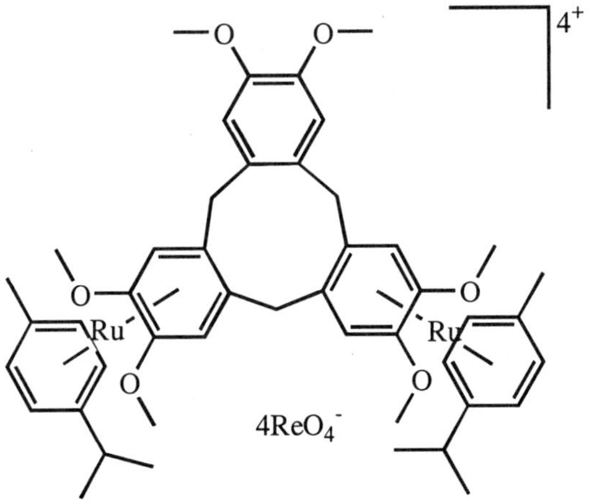

(**13**)

6.3 Neutral Calixarene Anion Receptors

Neutral anion receptors can be subdivided into two types, those relying solely on hydrogen bonds to bind anions, those containing electron deficient or Lewis acid centres. The latter group often also contain hydrogen bonding moieties. It is not only the nature of the receptor which affects the strength of anion binding but also the solvent in which the studies are carried out. In general binding studies for neutral receptors tend to be carried out in less polar, less competitive solvents than their charged counterparts and this needs to be taken into account when comparing the magnitude of reported stability constants.

6.3.1 Calixarene Based Hydrogen Bond Anion Receptors

Several different hydrogen bonding groups have been incorporated in upper and lower rim calixarene frameworks to create neutral anion receptors including amides, amines and alcohols.

Urea and thiourea upper[31, 32] and lower[33, 34] rim derivatised calixarenes have been synthesised. Reinhoudt *et al.* synthesised the calix[6]arene based receptor (**11**) and the analogous calix[4]arene based receptor which were both found to bind Cl⁻ and Br⁻ anions in $CDCl_3$ with 1:1 stoichiometry

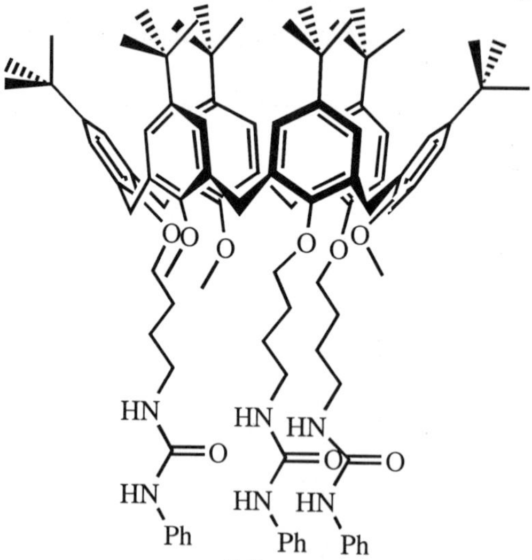

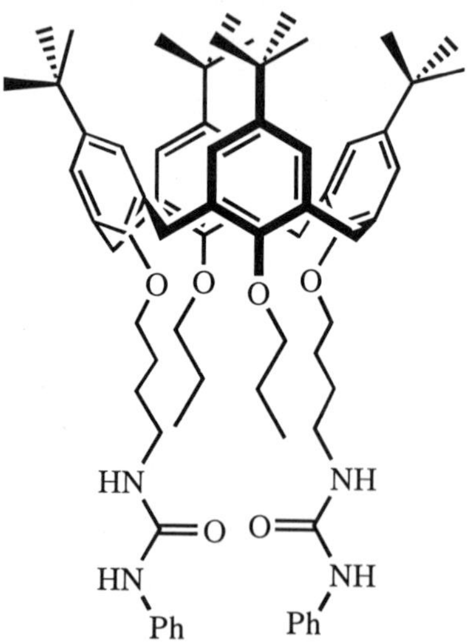

Table 6.7 Stability constants (M^{-1}) for receptors (**14**) and (**15**) in $CDCl_3$ determined by ^1H NMR titration. [a] Errors estimated ≤5%.

Anion	K^a (M^{-1})	
	(**14**)	(**15**)
Cl^-	480	7105
Br^-	1450	2555

Receptor (**14**) was found to show a preference for Br^- over Cl^- in $CDCl_3$ solution, whereas receptor (**15**) displayed the opposite selectivity. (**Table 6.7**) This suggests that the cavity formed by the three urea derivatives in receptor (**14**) is of a size that is more complementary to the larger Br^- anion than Cl^-, and that this effect is strong enough to outweigh the expected higher hydrogen bonding affinity of the d Cl^- anion for the urea hydrogens, whereas in receptor (**15**) both factors favour Cl^-. The tetra urea substituted analogue of (**15**) bound anions more weakly, possibly due to increased intramolecular hydrogen bonding competing with anion binding.

The thiourea analogue of (**14**) was found to bind halides more weakly but showed a strong affinity for the symmetric 1,3,5 benzenetricarboxylate anion[34]

The urea receptors use amide N-H bonds to form hydrogen bonds to anions. The strength of binding will obviously be related to the strength of the hydrogen bonds which is in turn related to the polarity of the bond. Receptors (**16a-c**) were synthesised by Loeb et. al. to examine the effect of electron withdrawing substituents α to the amide, increasing the amide N-H bond acidity, on the strength of anion binding[35]. Molecules (**17a-c**) were used as a non-calixarene comparison. (**Table 6.8**)

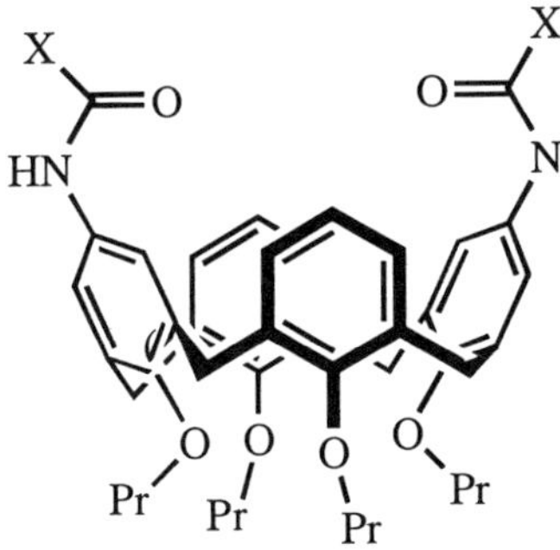

(**16a**) X = CH_2Cl
(**16b**) X = $CHCl_2$
(**16c**) X = CCl_3

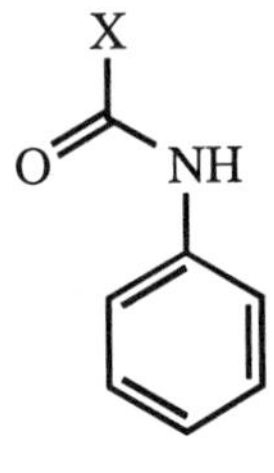

(**17a**) X = CH_2Cl
(**17b**) X = $CHCl_2$
(**17c**) X = CCl_3

Table 6.8 K (M^{-1}) for receptors (**16a-c**) and model compounds (**17a-c**) with benzoate anion in CDCl$_3$ determined by ^{1}H NMR titration. [a] No binding observed.

Receptor	16a	16b	16c	17a	17b	17c
K	107	5160	a	81	210	a

The addition of a second electron withdrawing group to the α–position of the amide results in an approximate 50 fold increase in binding for (**16b**) relative to (**16a**). A similar effect is seen for the model compound. The lack of binding for the tri-chloro compound is probably due to the steric effect of the large CCl$_3$ group preventing anion binding as it affects both (**16c**) and the analogous model compound (**17c**). The importance of the calixarene in preorganising two amide functions in a conformation suitable for binding is demonstrated by the much higher binding strength of the calixarene based receptor compared to the model. In the calixarene receptor two amides N-H bonds can co-operate in binding an anion with 1:1 stoichiometry. In the model compound only one amide is available for 1:1 binding so the resultant binding is much weaker.

Reinhoudt *et al.* synthesised the upper rim functionalised sulphonamide receptors (**18a-c**)[36] which contain electron withdrawing SO$_2$ groups, resulting in increased polarity of the N-H bonds and stronger hydrogen bonds with anions. They all display a remarkable selectivity for hydrogen sulphate over other anions in CDCl$_3$. Receptor (**18c**) shows the strongest binding due to the additional amide functions. (**Table 6.9**)

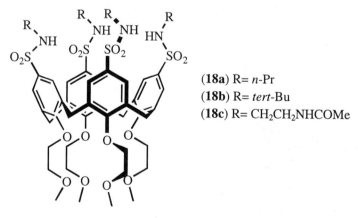

(**18a**) R= *n*-Pr
(**18b**) R= *tert*-Bu
(**18c**) R= CH$_2$CH$_2$NHCOMe

Table 6.9 K (M^{-1}) for receptors (**18a-c**) in CDCl$_3$ determined by ^{1}H NMR titration. Errors estimated <5%. ^{a}Counterion is Bu$_4$N$^+$ ^{b}shows complicated binding, no K value for 1:1 binding could be determined.

Receptor	Anion^a			
	H$_2$PO$_4^-$	HSO$_4^-$	Cl$^-$	NO$_3^-$
18a	350	970	360	240
18b	<10	134	72	43
18c	b	103400	1250	513

Beer *et al.* have synthesised the upper rim to lower rim linked bis calix[4]arene receptor (**19**)^{37}. Addition of anions to receptor (**19**) resulted in significant perturbations of the amide, upper calixarene phenolic and lower calixarene aromatic protons indicating that the binding site is between the two calixarenes.

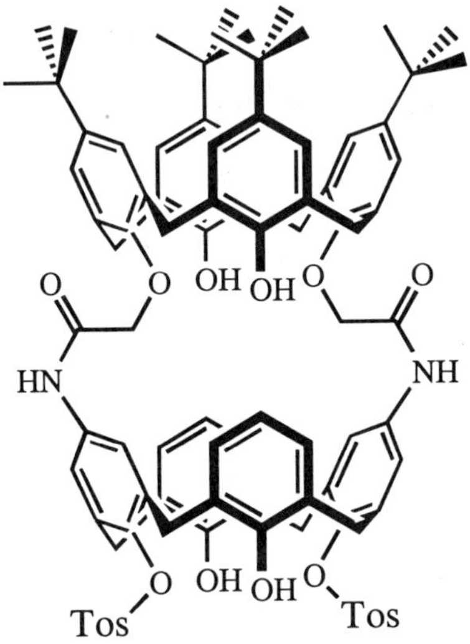

(19)

Anion binding studies were carried out by ^{1}H NMR in CD$_2$Cl$_2$ and receptor (**19**) was found to bind anions with a high selectivity for fluoride. (**Table 6.10**) The selectivity can be ascribed to the size of the anions compared to the calixarene cavity. Molecular modelling calculations suggest that the cavity is too small to encapsulate either HSO$_4^-$ or H$_2$PO$_4^-$.

Table 6.10 K (M^{-1}) for receptors (**19**) in CD$_2$Cl2 determined by ^{1}H NMR titration. ^{a}Errors estimated < 10%.

Anion	K^a (M^{-1})
F$^-$	1330
Cl$^-$	172
HSO$_4^-$	21
H$_2$PO$_4^-$	91

Ungaro *et al.* have recently utilised the electron withdrawing trifluoromethyl moiety to increase the polarity, and thus hydrogen bonding potential of a O-H bond in receptor (**20**)[38]

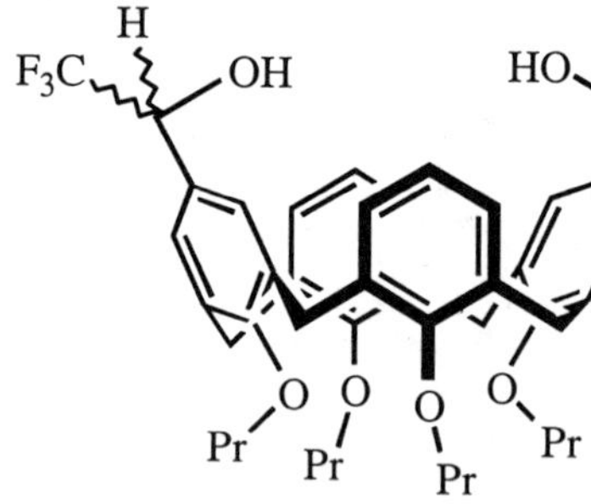

(**20**)

Receptor (**20**) shows a preference for carboxylate over spherical anions in CDCl$_3$ as has been found with other upper rim di substituted calix[4]arenes. A tetra-substituted analogue with C$_6$F$_5$ in the place of CF$_3$ showed a preference for halides. Interestingly the racemic (RR, SS) form of (**20**) was found to bind acetate more strongly then the *meso* (SR, RS) form, possibly due to unfavourable steric interactions in the *meso* form.

6.3.2 Calixpyrroles

The term calixpyrrole was first coined by Sessler[39] to describe octaalkyl porphyrinogens e.g. (**21**) that may be regarded as heterocalixarene analogues. Whereas calixarenes are the cyclic oligomers of phenols and formaldehyde, calixpyrroles are the cyclic oligomers of pyrroles and ketones.

(21) R ≠ H, R groups not necessarily identical

Previously it has been shown that functionalised calixarenes will bind anions, unmodified calixarenes show no such interaction. Calixpyrroles will however bind anions in weakly co-ordinating solvents without further modification. The anion interacts with the N-H bonds of the pyrroles forming strong hydrogen bonds. In the uncomplexed state the calixpyrrole exists in the 1,3 alternate form. Upon complexation to anions it adopts the full cone conformation allowing all four hydrogens to be involved in binding (**Figure 6.3**)

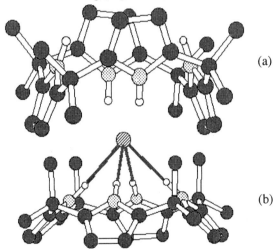

Figure 6.3 Molecular structures of **(21)** where R=Me (a) without anion present (b) as the chloride complex.

Calixpyrroles exhibit strong anion binding, with the simple members of the group showing a strong preference for fluoride anions in CD_2Cl_2. Substitution at the carbon rim of calixpyrroles has been used to 'tune' the anion binding properties of the receptors[40]. The presence of electron donating groups reduced the strength of binding relative to the unsubstituted analogue whereas electron withdrawing groups enhanced binding, the different groups effecting the acidity of the N-H protons and in turn their binding ability. Further modifications of the basic calixpyrrole structure have been carried out. A calixpyrrole with a ferrocene moiety attached via a linking group has been synthesised, and the electrochemical properties have been examined[41]. The electrochemical behaviour was found to be affected by anion complexation. Calixpyrroles have also been attached to silica gels and used for the HPLC separation of anionic nucleotides and oligonucleotides and neutral fluorinated biphenyls[42].

6.3.3 Neutral Lewis Acid Metal Containing Receptors.

A large number of neutral calixarene based receptors have been synthesised containing Lewis acid centre metal containing fragments that act as electron deficient centres. Receptors of this type can act in two ways.

(a) The anion donates electron density directly to vacant orbitals on the Lewis acid centre forming a weak bond as in the non calixarene boron receptor (**6**) described earlier.

(b) The electron deficient/electropositive centre withdraws electron density from surrounding groups which can then interact with the anion. This strategy is commonly used in conjunction with hydrogen bonding groups, e.g. the electron withdrawing group is used to enhance the acidity of amide N-H bonds.

Several of the receptors that act in the second manner are direct analogues of the positively charged receptors described earlier in this chapter. e.g. the analogues of receptors (**9a**) and (**9b**) where R=Re(bipy)(CO)$_3$Cl have been synthesised[25] by Beer et. al. with the rhenium(I) metal withdrawing electron density to enhance the acidity of the amide protons. The strength of binding of these rhenium receptors is relatively weaker than with the charged ruthenium(II) analogues as there is no electrostatic attraction e.g. K=4400 and 2550 M^{-1} for the Ru(bipy)$_3^{2+}$ and Re(bipy)CO$_3$Cl analogues of (**9a**) with H$_2$PO$_4^-$ respectively in d_6-DMSO. However the nature of the binding interactions with the amide protons and binding geometry are similar and as a consequence exhibit similar selectivity trends.

Solvent can have a dramatic effect on the strength and selectivity of anion binding receptors, and as noted earlier, this is especially important when comparing neutral and charged receptors where binding studies are often carried out in different solvents due to differing solubilities. Recently Beer *et al.* carried out anion binding studies in a variety of solvents using the bis calixarene ferrocene receptor (22)[9]. The calixarene is important in helping the receptor to be soluble in a variety of solvents, as well as in determining receptor selectivity. ^{1}H NMR titration experiments were carried out with receptor (22) in CD$_2$Cl$_2$, CD$_3$CN and (CD$_3$)$_2$CO solutions with tetrabutylammonium chloride, benzoate and acetate salts. The results are summarised in **Table 6.11**

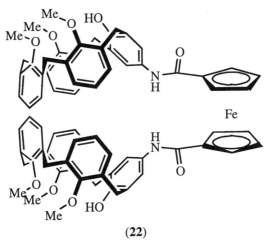

(22)

Table 6.11 Solvent effects on the stability constant values for receptor (22) [a]Errors estimated < 10%, ε = relative permittivity, μ = dipole moment, AN = acceptor number of solvent.

Solvent	ε	μ	AN	K^a (M^{-1}) Anion		
				Cl$^-$	PhCO$_2^-$	MeCO$_2^-$
CD$_2$Cl$_2$	8.9	1.5	20.4	40	117	26
CD$_3$CN	36.0	3.96	18.9	70	360	120
(CD$_3$)$_2$CO	20.7	2.86	12.5	5200	940	6000

The solvent exerts a large effect on the equilibrium between the receptor and anionic guest, with changes of two orders of magnitude in some cases. There is no obvious relationship between the strength of binding and relative permittivity or dipole moment of the solvent. There is however an apparent correlation between K and the Guttman acceptor number (AN)[43] of the solvent. The acceptor number gives a quantitative measure of the solvent hydrogen-bond donor ability, a solvent with a large AN acts as a more effective hydrogen bond donor and so solvates anions to a greater extent competing with receptor binding. As the solvent AN number decreases the selectivity of the receptor changes, the smaller harder anions Cl⁻ and $MeCO_2^-$ become bound preferentially over the larger charge diffuse benzoate. The more heavily solvated anions are more greatly affected by changes in solvent, binding more strongly to the receptor, the best hydrogen bonding source available, in the lower AN solvents. The selectivity change observed on changing solvents is amplified by the calixarene receptor when compared to a non-calixarene model system. In order to better understand these trends an examination of the enthalpic and entropic contributions to binding would be useful. De Namor *et al.* have carried out some preliminary studies on the thermodynamics of other calixarene based anion receptors[44].

6.3.4 Calixresorcinarene Phosphine Complexes

In these receptors synthesised by Puddephatt *et al.* the calix[4]resorcinarene is substituted at the upper rim with organophosphine moieties which can in turn be complexed to metal ions e.g. Cu(I), Ag(I)[45]. Receptor (**23**) has a chloride anion trapped in the calix bowl co-ordinated to three of the copper atoms with longer, apparently weaker bonds than the bridging chloride atoms. The anion is fluxional between the four copper atoms. It is probably the electrophilic, Lewis acid nature of the copper atoms around the rim of the calixresocinarene that attract the anion. Binding studies indicate that there is selectivity for iodide over chloride probably due to the larger size of the iodide anion allowing it to co-ordinate to all four copper atoms rather than just three. This preference for iodide has been shown to enhance the rate of the substitution reaction of alkyl iodides by chloride in the presence of the silver analogue of (**23**) in CD_2Cl_2. During the reaction the chloride anion in the cavity is exchanged for the iodide of the alkyl halide.

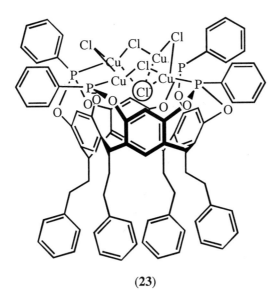

(23)

6.3.5 Ditopic Calixarene Receptors for Anion Recognition

The simultaneous binding of an anion and its counter cation, ion pair recognition, in a neutral bifunctional receptor utilises the positive charge on the cation to enhance the strength of anion binding. This area has recently attracted increasing attention, it utilises all the various strategies discussed so far, often combining them within one molecule. Receptors use combinations of electron deficient centres, hydrogen bonding and the electrostatic attraction of the counter cation to enhance anion binding. Ion pair binding has many possible commercial applications such as improving the extraction ability of receptors for environmentally important anions.

One of the earliest calixarene containing examples of this type was receptor (**24**) prepared by Beer et. al.[46]. Two benzo-15-crown-5 groups are attached to the lower rim of calix[4]arene via an amide linker. In the absence of an alkali metal cation receptor (**24**) showed very little interaction with anions in CD_3CN. Upon addition of potassium ions a sandwich complex is formed between the two benzo-crown ethers. This results in the amide groups being held close together, and along with the electrostatic attraction of the alkali metal cation, preorganises (**24**) to bind anions Dihydrogen phosphate is particularly strongly bound $K = >10^4$ in CD_3CN solution.

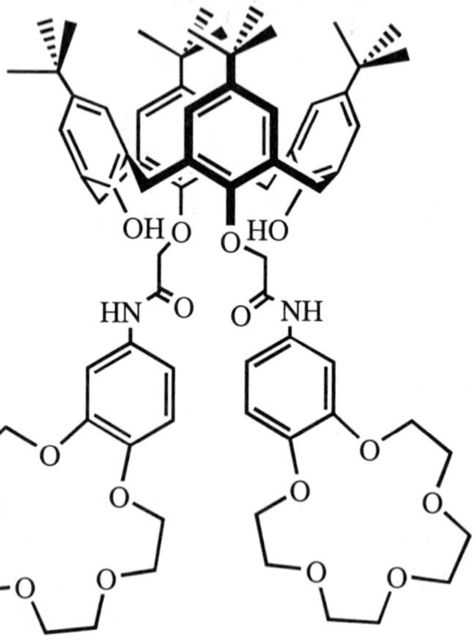

(**24**)

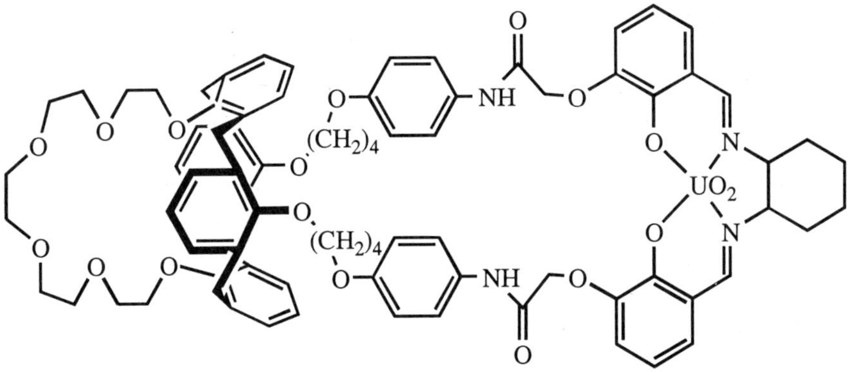

(25)

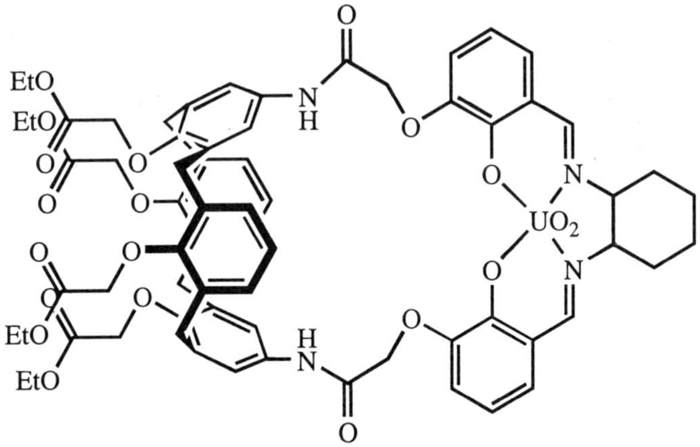

(26)

The bifunctional receptors (25)[47] and (26)[48] synthesised by Reinhoudt *et. al.* use a combination of amide hydrogen bonding and the Lewis acid uranyl centre for anion binding. A cation can be simultaneously co-bound by the lower rim crown and tetraester respectively.

Receptor (25) was found to transport hydrophilic CsCl through a supported liquid membrane, while a model receptor containing only the anion binding site shows very little transport effect.

Receptor (26) was found to selectively bind $H_2PO_4^-$ anions in d_6-DMSO with K=390 M^{-1}, other anions such as Cl⁻ showed little interaction. The receptor has a tetraester substituted lower rim which has been shown to selectively bind sodium cations[49]. After stirring receptor (26) and NaH_2PO_4 in MeCN-H₂O, 10:1, the positive FAB mass spectrum contained an intense peak corresponding to [(**26**) + Na$^+$]$^+$ the 1:1 complex while the corresponding negative FAB mass spectrum of the same sample contained a peak corresponding to [(**26**) + H$_2$PO$_4$$^-$]$^-$ demonstrating the complexation of both cation and anion in the same bifunctional molecule.

In the case of receptor (**24**) the addition of alkali metal cations switches on anion binding by preorganising the receptor. Another example of this effect is seen with receptor (**27**) and its tetra upper-rim substituted analogue[50].
In the absence of sodium ions the receptor is hydrogen bonded across the upper rim preventing anion binding in CDCl$_3$ solution.

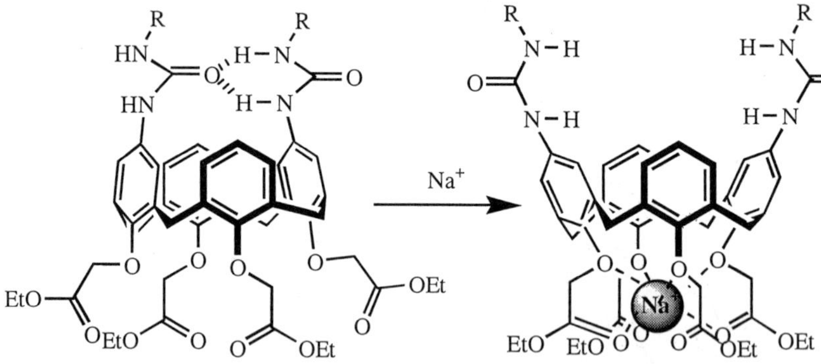

Figure 6.4 Complexation of receptor (**27**) with sodium ions breaking the hydrogen bonds across the upper rim.

Complexation of sodium ions on the lower ester functionalised rim of the calixarene pulls the upper rim apart. (**Figure 6.4**) This has the effect of freeing the amide N-H groups and allowing anion binding. The tetra substituted analogue of (**27**) behaves in a similar manner. In the absence of cations no interaction of the urea protons with anions was observed. In the presence of a co-bound sodium ion the receptor binds both chloride (K= 1 x 10^4 M^{-1}), and bromide (K= 1.3 x 10^3 M^{-1}) in CDCl$_3$ solution.

Related receptors (**28a-b**) have recently been synthesised using the more acidic thio urea on the upper rim of the calixarene[32]. The presence of the single binding group prevents intramolecular hydrogen bonding and allows the strength of anion binding both with and without a co-bound cation to be examined.

When there was a methylene group between the aromatic carbon and the thiourea unit (**28a**) it was found that the strength of anion binding was little effected or decreased in the presence of a co-bound cation, whereas when the thiourea was bound directly to the aromatic group a large enhancement of binding occurred in most cases in the presence of a co-bound cation. (**Table 6.12**)

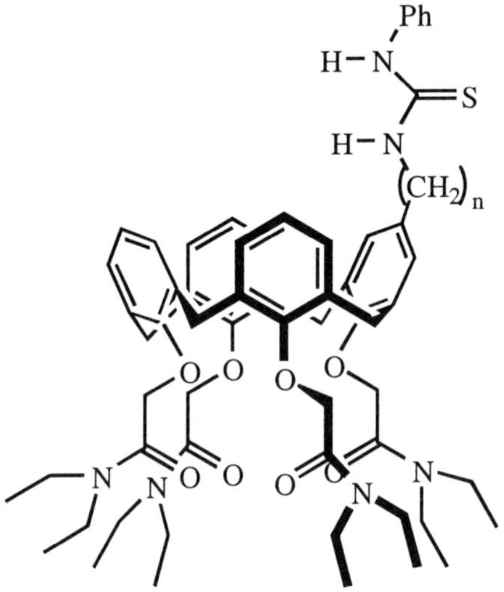

(**28a**) n=1
(**28b**) n=0

Table 6.12 K (M^{-1}) for receptors (**28a-b**) in d_6-DMSO with selected anions determined by ^{1}H NMR titration. Errors ±10%. ^{a}Counterion is Bu$_4$N$^+$

	Host			
Anion	**28a**	[Na(**28a**)]$^+$	**28b**	[Na(**28b**)]$^+$
Benzoate	175	190	250	1100
Acetate	470	330	940	1200
Propionate	280	215	250	1000

The presence of the methylene link in receptor (**28a**) prevents the thiourea amides from feeling the electron withdrawing effect of the co-bound cation on the lower rim. In receptor (**28b**) the direct link to the aromatic nucleus allows withdrawal of electron density from the thiourea and thus enhanced hydrogen bonding and anion binding ability.

One problem with the study of anion binding enhancement by the presence of a co-bound cation is that there are a number of equilibria involved. The most important of these is the equilibrium between the ion pair bound in the receptor, ion pairing in solution and precipitation of the salt itself. If care is taken to ensure that all components are soluble in the chosen solvent then ion pairing in solution competing with ion pair binding in the receptor is the most important factor.

In an attempt to investigate this quantitatively Beer *et. al.* synthesised receptors (**29**) and (**30**)[51]. ^{1}H NMR titrations were carried out using metal iodides, with the cation added as the ClO_4^- or PF_6^- salt and the anion as the NBu_4^+ salt all of which were soluble in the chosen solvent acetonitrile.

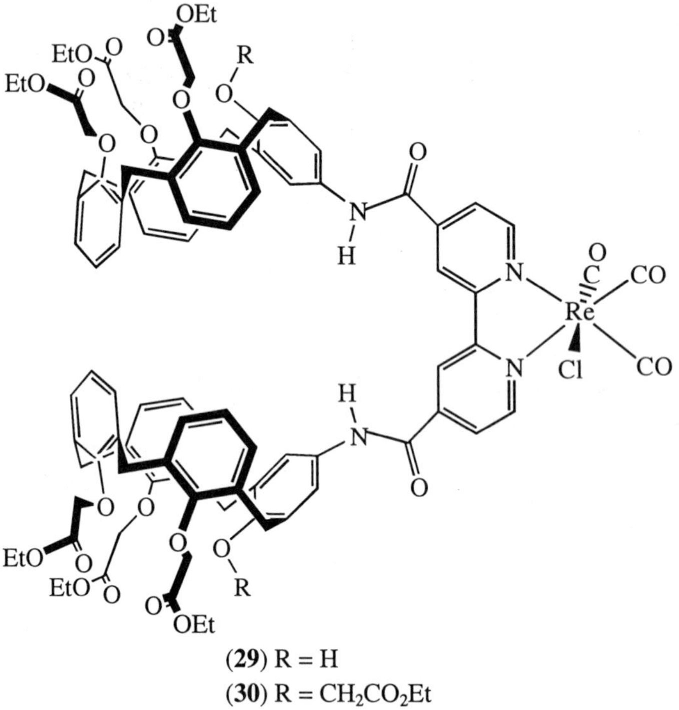

(**29**) R = H
(**30**) R = CH_2CO_2Et

The results of binding studies of (**29**) and (**30**) in acetonitrile with iodide and iodide in the presence of two equivalents of co-bound metal cation (Li$^+$, Na$^+$ and K$^+$) are shown in **Figure 6.5**

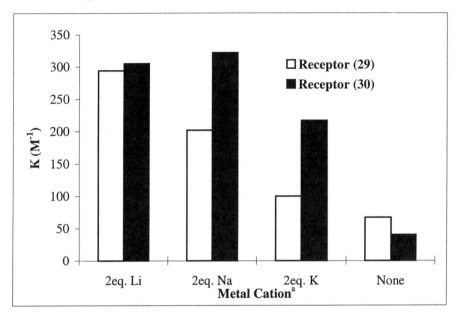

Figure 6.5 K (M^{-1}) for receptors (**29**) and (**30**) with iodide anions alone and iodide anions in the presence of two equivalents of lithium, sodium and potassium cations. [a] Li and Na added as perchlorate, K as hexafluorophosphate.

With both receptors it can be seen that there is a significant increase in the strength of anion binding in the presence of the co-bound metal ions compared to the free receptor, almost ten fold in the case of (**30**) with co-bound Na$^+$ ions. This enhancement is probably due to a combination of effects, the enhancement of amide acidity by the electron with drawing metal, preorganisation of the calixarene for binding by the metal rigidifying the structure and of course electrostatic attraction. For receptor (**29**) the binding strength in the presence of the metal decreases evenly on going from Li$^+$ to Na$^+$ to K$^+$, whereas for receptor (**30**) there is an alternation about Na$^+$. Ion pairing in solution competes with ion pair binding in the receptor, and this can be used to explain the observed trends. The trend that might be expected for ion pairing in solution is KI > NaI > LiI. For receptor (**30**) metal binding is weak, ion pairing competes effectively and the binding strength decreases with increasing competition from ion pairing giving the expected trend of K LiI > NaI > KI. For receptor (**30**) the trend is interrupted for NaI due to the known high selectivity of the lower rim tetra-ester substituted calixarene for Na^+ ions[49].

6.4 Conclusion

In this chapter we have reviewed the different ways that calixarenes have been used to make selective anion receptors. The calixarene framework plays an important role in developing these selectivity trends that are often different from those of non-calixarene analogues. Topologically the three dimensional calixarene structure allows for the design of binding sites that are complementary to a target anion. The ease of modification of both the upper and lower rims allows for the building of a wide range of structures both for binding and to enhance receptor solubility the latter being of great importance for extraction and membrane transport applications. Some of the receptors described have potential for commercial application, as extractants for anionic pollutants or in sensing membranes, while others have been developed to further elucidate the fundamental aspects of anion co-ordination chemistry. In all cases they show the versatility of calixarenes as platforms on which to build anion receptors and their ever increasing importance in the field of anion co-ordination chemistry.

6.5 References

1. Dietrich B., *Pure Appl. Chem.*, **65**, 1993, 1457.
2. Bianchi A., Bowman-James K., García-España E. (1997) Supramolecular Chemistry of Anions. Wiley-VCH, New York.
3. Harrison R.M. (1983) *Pollution: Causes Effects and Control*. The Royal Society of Chemistry, London.
4. Quinton P.M., *FASEB J.*, **4**, 1990, 2709.
5. Simmons H.E., Park C.H., *J. Am. Chem. Soc.*, **90**, 1968, 243.
6. Pedersen C.J., *J. Am. Chem. Soc.*, **89**, 1967, 7017.
7. Fraústo da Silva J.J.R., Williams R.J.P. The Biological Chemistry of the Elements, (1991) Clarendon Press, Oxford.
8. Greenwood N.N., Earnshaw A. (1984) *Chemistry of the Elements*. Pergamon Press, Oxford.
9. Beer P.D., Shade M., *Chem. Commun.*, 1997, 2377.
10. Chakrabarti P., *J. Mol. Biol.*, **234**, 1993, 463 and references within.
11. Luecke H., Quiocho F.A., *Nature*, **347**, 1990, 402.
12. Berger M., Schmidtchen F.P., *Chem. Rev.*, **97**, 1997, 1609.
13. Graf E., Lehn J.-M., *J. Am. Chem. Soc.*, **98**, 1976, 6403.
14. Dietrich B., Guilhem J., Lehn J.-M., Pascard C., Sonveaux E., *Helv. Chem. Acta.*, **67**, 1984, 91.
15. Schmidtchen F.P., *Angew. Chem. Int. Ed. Eng.*, **16**, 1977, 720.
16. Dietrich B., Fyles D.L., Fyles T.M., Lehn J.-M., *Helv. Chim. Acta.*, **62**, 1979, 2763.
17. Serpersu E.H., Shortle D., Mildvan A.S., *Biochemistry*, **26**, 1987, 1289.
18. Newcomb M., Madonik A.M., Blanda M.T., Judice J.K., *Organometallics*, **6**, 1987, 145.
19. Jung M.E., Xiu H., *Tetrahedron Lett.*, **29**, 1988, 297.
20. Ogawa K., Aoyagi S., Takeuchi Y., *J. Chem. Soc. Perkin Trans. 2*, **2**, 1993, 2389.
21. Beauchamp A.L., Oliver M.J., Wuest J.D., Zacherie B., *J. Am. Chem. Soc.*, **108**, 1986, 73.

22. Reetz M.T., Niemeyer C.M., Harms K., *Angew. Chem. Int. Ed. Engl.*, **30**, 1991, 1472.
23. Beer P.D., Drew M.G.B., Hazlewood C., Hesek D., Hodacova J., Stokes S.E., *J. Chem. Soc., Chem. Commun.*, 1993, 229.
24. Beer P.D., Drew M.G.B., Hesek D., Nam K.C., *Chem. Commun.*, 1997, 107.
25. Beer P.D., Drew M.G.B., Hesek D., Shade M., Szemes F., *Chem. Commun.*, 1996, 2161.
26. Beer P.D., Szemes F., Hesek D., Chen Z., Grieve A., Goulden A.J., Wear T., *J. Chem. Soc. Chem. Commun.*, 1994, 1269.
27. Beer P.D., Szemes F., Hesek D., Chen Z., Dent S.W., Drew M.G.B., Goulden A.J., Mortimer R.J., Wear T., Weightman J.S., *Inorg. Chem .*, **35**, 1996, 5868.
28. Beer P.D., Shade M., *Gazz. Chim. Ital.*, **127**, 1997, 651.
29. Atwood J.L., Steed J.W., Hancock K.S.B., Holman K.T., Staffilani M., Juneja R.K., Burkhalter R.S., *J. Am. Chem. Soc.*, **119**, 1997, 6324.
30. Steed J.W., Juneja R.K., Atwood J.L., *Angew. Chem. Int. Ed. Engl.*, **33**, 1994, 2456.
31. Casnati A., Fochi M., Minari P., Pochini A., Reggiani M., Ungaro R., Reinhoudt D.N., *Gazz. Chim. Ital.*, **126**, 1996, 99.
32. Pelizzi N., Casnati A., Friggeri A., Ungaro R., *J. Chem. Soc. Perkin Trans. 2.*, 1998, 1307.
33. Scheerder J., Fochi M., Engbersen J.F.J., Reinhoudt D.N., *J. Org. Chem.*, **59**, 1994, 7815.
34. Scheerder J., Engbersen J.F.J., Casnati A., Ungaro R., Reinhoudt D.N., *J. Org. Chem.*, **60**, 1995, 6448.
35. Cameron B.R., Loeb S.J., *Chem. Commun.*, 1997, 573.
36. Morzherin Y., Rudkevich D.M., Verboom W., Reinhoudt D.N., *J. Org. Chem.*, **58**, 1993, 7602.
37. Beer P.D., Gale P.A., Hesek D., *Tetrahedron Lett.*, **36**, 1995, 767.
38. Pelizzi N., Casnati A., Ungaro R., *Chem. Commun.*, 1998, 2607.
39. Gale P.A., Sessler J.L., Král V., Lynch V., *J. Am. Chem. Soc.*, **118**, 1996, 5140.

40. Gale P.A., Sessler J.L., Allen W.E., Tvermoes N.A., Lynch V., *Chem. Commun.*, 1997, 665.
41. Sessler J.L., Gebauer A., Gale P.A., *Gazz. Chim. Ital.*, **127**, 1997, 723.
42. Sessler J.L., Gale P.A., Genge J.W., *Chem. Eur. J.*, **4**, 1998, 1095.
43. Mayer U., Gutmann V., Gerger W., *Monatsh. Chem.*, **106**, 1975, 1235.
44. Danil de Namor A.F., Hutcherson R.G., Verlade F.J.S., Ormachea M.L.Z., Salazar L.E.P., Jammaz I.A., Rawi N.A., *Pure Appl. Chem.*, **70**, 1998, 769.
45. Xu W., Vittal J.J., Puddephatt J., *J. Am. Chem. Soc.*, **117**, 1995, 8362.
46. Beer P.D., Drew M.G.B., Knubley R.J., Ogden M.I., *J. Chem. Soc. Dalton Trans.*, 1995, 3117.
47. Rudkevich D.M., Mercer-Chalmers J.D., Verboom W., Ungaro R., de Jong F., Reinhoudt D.N., *J. Am. Chem. Soc.*, **117**, 1995, 6124.
48. Rudkevich D.M., Verboom W., Reinhoudt D.N., *J. Org. Chem.*, **59**, 1994, 3683.
49. Arnaud-Neu F., Collins E.M., Deasy M., Ferguson G., Harris S.J., Katiner B., Lough A.J., McKervey M.A., Marques E., Ruhl B.L., Schwing-Weill M.J., Seward E.M., *J. Am. Chem. Soc.*, **111**, 1989, 8681.
50. Scheerder, J., v. Duyhoven J.P.M., Engbersen J.F.J., Reinhoudt D.N., *Angew. Chem. Int. Ed. Engl.*, **35**, 1996, 1090.
51. Beer P.D., Cooper J.B., *Chem. Commun.*, 1998, 129.

CHAPTER 7

STRUCTURAL PROPERTIES AND THEORETICAL INVESTIGATION OF SOLID STATE CALIXARENES AND THEIR INCLUSION COMPLEXES.

F. UGOZZOLI.
Dipartimento di Chimica Generale ed Inorganica Chimica Analitica Chimica Fisica, Università di Parma, Parco Area delle Scienze 17/a, I-43100 Parma, Italy.

7.1. Introduction.

Although several hundreds of crystal structures of calixarenes are now known, only the crystallographic data of about 600 of them have been deposited and stored in the Cambridge Crystallographic Data File, so that a complete and systematic study on the structural properties in the solid state is prevented.

Thus this chapter will not deal with a simple, although exhaustive, listing of the structural properties of calixarenes in the solid state, but it will try to focus on the action of specific weak non-covalent intermolecular forces and on the way in which they can be exploited for crystal engineering.

7.2. The $CH_3...\pi$ interaction.

Several solid state structures demonstrate that the intramolecular cavities of calix[4]arenes in the *cone* conformation are able to guest aromatic neutral molecules giving highly stable crystals.

Among them: a) the structure of the toluene ⊂ **1b** complex [1], since it showed for the first time the peculiar ability of calixarenes to give inclusion complexes (this structure was refined to higher resolution in 1996 [2], b) the structure of benzene ⊂ **1b** complex [3] which is isostructural with that of the toluene complex, c) the structure of the nitrobenzene ⊂ **1b** complex [4], d) the p-xylene ⊂ **1c** complex [5], e) the toluene ⊂ **1c** complex [6]. But also 2:1, or cage type, host-guest complexes have been observed: *i.e.* f) the structure of the 1:2 anisole ⊂ **1b** [7].

The guest selectivity properties of **1a** and **1c** toward aromatic molecules obtained by competitive crystallization experiments in the presence of equal volumes of competing guests [8] showed that the *cone* conformation of the host is not a sufficient condition to ensure the selective complexation of the aromatic guest inside the intramolecular cavity of the host, but that other structural factors become predominant. This fact, together with the experimental proof that p-H calix[4]arenes never gave complexation, led to the hypothesis of an attractive $CH_3...\pi$ interaction between the p-*tert*-butyl groups at the *upper-rim* of the hosts and the aromatic nuclei of the guests.

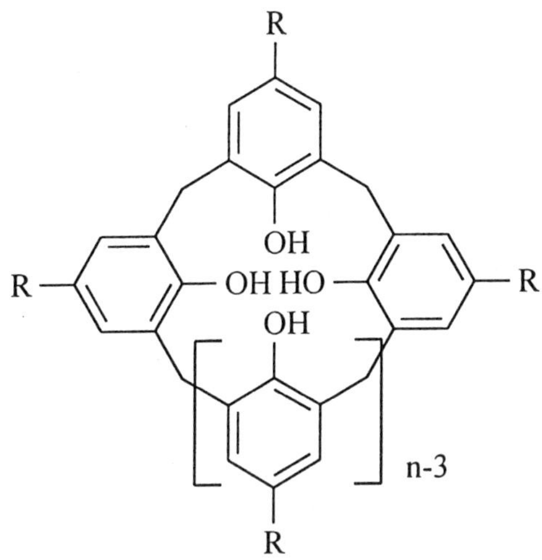

1a: n=4 R=H
1b: n=4 R=Bu^t
1c: n=4 R=i-Pr
1d: n=4 R=C$_8$H$_{17}$
1e: n=5 R=Bu^t
1f: n=6 R=H

Other structural evidence has enforced this hypothesis: *i.e.* the structure of the pyridine ⊂ 1,3-dihydroxy-p-*tert*-butylcalix[4]arene-crown-6 (**2**) complex [9] showed that the pyridine molecule is held within the calixarene in close contact with two opposite *tert*-butyl groups, thus indicating that complexation of one aromatic molecule inside the cavity is scarcely influenced by the presence of *upper-rim* substituents.

A more detailed analysis of the host-guest interactions has been done by molecular mechanics calculations carried out starting from the atomic co-ordinates obtained from the crystallographic study. In our approach both the host and the guest were treated as rigid bodies and the total potential energy of the system was calculated as a function of any rigid rotation of the guest. The calculations demonstrated that: a) the presence of only the van der Waals (U_{vdw}) and electrostatic (U_{el}) contributions cannot predict the ordered structure observed. b) another type of host-guest interaction has to be invoked to explain the complexation of the pyridine. c) A conclusive result was obtained by adding to the total potential energy a further contribution coming from the CH$_3$...π interaction which is described by an effective "Morse-like" potential U_{eff} given by the equations 1.1 and 1.2.

$$U_{eff} = D[1-\exp(\ln 2(r-r_0)/(r_1-r_0))]^2 - D \qquad \text{for } r<R \qquad (1.1)$$

$$U_{eff} = U_{vdw} + U_{el} \qquad \text{for } r>R \qquad (1.2)$$

Where the centres of force are the donor (each H atom of the host p-*tert*-butyl groups) and the acceptor (each C atom of the guest) atoms separated by a distance r; r_0 (2.71 Å) is the abscissa of the minimum $U_{eff} = -D$. For values $r < r_1$ (2.68 Å) the interaction becomes repulsive.

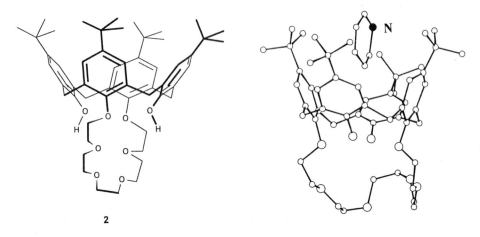

Figure 7.1. X-ray crystal structure of 1:1 complex between 1,3-dihydroxy-p-*tert*- butylcalix[4]arene-crown-6 and pyridine (**2**).

The R value (3.5 Å) is imposed by the continuity condition between the two functions. The value of D (0.1 kcal mol^{-1}) has been estimated from the literature [10]. With the contribution due to the $CH_3...\pi$ the pattern of the total potential energy completely predicted the structure observed in the solid state.

7.2.1. Almost Free Methyl Quantum Rotors in Calixarenes.

In inclusion complexes the hindering barrier provided by the host molecules to the reorientation of the guests can be accurately obtained by Inelastic Neutron Scattering (INS) measurements. In such experiments the orientational freedom of the guest is exploited as a probe for the measurements of the host-guest potential energy barrier.

In particular, the low-energy dynamics of methyl groups in a solid exhibit specific characteristics arising from the indistinguishability of the protons due to the *tunnel effect* [11]. Thus, in principle, INS measurements on inclusion complexes with guests supporting methyl groups are good candidates for the accurate study of the host-guest interactions.

INS measurements have been carried out to explore the tunnelling of methyl groups belonging to several guest molecules encapsulated in the intramolecular cavity of the p-*tert*-butylcalix[4]arene host **1b** molecule with the guest CH_3 group oriented near the *lower rim* as shown by the X-ray diffraction studies: toluene [12], p-xylene [13], γ-picoline [14]-[15].

In all the cases investigated, the low temperature neutron spectra show a number of bands which may be interpreted as being due to transitions between tunnel-split librational states of the methyl groups. The main line occurs near 0.63 meV, very close to the CH_3 quantum free rotor limit (0.655 meV). These results clearly indicate that the energy barrier provided by the host molecule is negligible and that the guest is prevented from filling efficiently the host cavity. Moreover, this suggests that the van der Waals and the electrostatic forces are not the only driving forces for the complexation of the guest but that other factors co-operate to hold the guest near to the *upper rim* of the calixarene than those one would expect. This is another indication that supports the CH_3...π attractive interaction between the methyl groups of the hosts and the aromatic moieties of the guests as a driving force for the inclusion of aromatic guests within the cavity of p-alkylcalix[4]arene derivatives.

7.3. The Cation...π interaction in action.

Several papers have demonstrated that the cation...π interaction often plays a fundamental rôle in molecular recognition, [16]-[21], enzyme active sites [22], ion channels [23] etc.

Calixarenes, and in particular calix[4]arenes, which can be synthesised in a variety of shapes by immobilisation of the conformational isomers, possess a π-basic surface originated by the four benzene rings; thus they provide a variety of molecular architectures useful for evaluating the contribution of the cation...π interaction in the complexation processes.

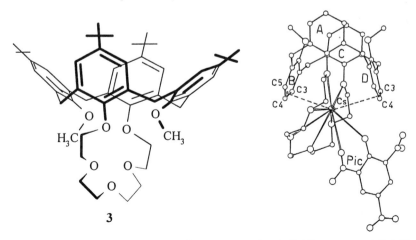

Figure 7.2. The molecular structure of the dimethoxy-p-*tert*-butylcalix[4]arene-crown-5 (3)CsPic (Pic=picrate) complex.

The first evidence of the weak cation...π interaction in calix[4]arene complexes has been found in the structural studies of the 1,3-dimethoxycalix[4]arene-crown-6 (3) Cs-Picrate complex [24] illustrated in Fig. 7.2 and of the 1,3-diisopropoxy calix[4]arene-crown-6 (4) Cs-Picrate complex [25] shown in Fig. 7.3.

Although the ligand (3) is conformationally mobile and exists in solution mainly in the *cone* conformation, in the solid state the complex shows two crystallographically independent molecules both in the *1,3 alternate* conformation in which the Cs^+ ion interacts in η^3 fashion with the C(3B), C(4B) and C(5B) atoms, and in η^2 fashion with the C(3D), C(4D) atoms of the two reversed arene rings. The Cs-C distances range from 3.431(10) to 3.684(8) Å in one symmetry independent complex unit and from 3.354(7) to 3.588(8) Å in the second one. The shortest Cs-C distances observed compare with those observed in systems for which the Cs^+...arene interactions have been documented [26]-[27].

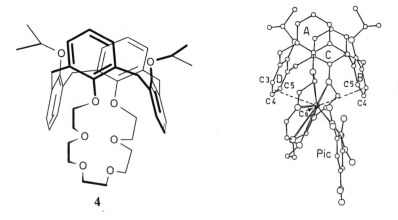

Figure 7.3. The molecular structure of the cationic [1,3-diisopropylcalix[4]arene-crown-6 (4) Cs]⁺ complex.

Also in the molecular structure of the 4•CsPicrate complex there is clear evidence for the participation of the Cs^+...π interaction in the complexation. As shown in Fig. 7.3. the co-ordination sphere at the Cs^+ cation is now quite different from that observed in the picrate complex of ligand **3** . However, although the wrapping mode of the crown moiety is quite different and in the complex of ligand **4** the picrate anion does not co-ordinate the metal cation, the Cs^+...π interaction is very similar: the Cs^+ cation interacts in η^3 fashion with the C(3D), C(4D) and C(5D) atoms , and in η^2 fashion with the C(4B) and C(5B) atoms of the two inverted nuclei B and D.

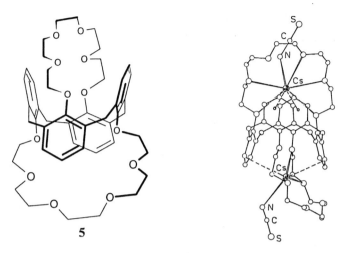

Figure 7.4. Perspective view of the dinuclear Cs$_2$(NCS)$_2$ complex of ligand **5**.

Also the Cs-C contacts, which range from 3.486(8) to 3.69(1) Å, are fairly comparable with the above cited values found in the picrate complex of the ligand **3**.

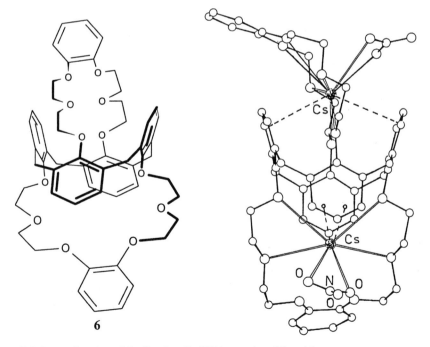

Figure 7.5. Perspective view of the dinuclear Cs$_2$(NO$_3$)$_2$ complex of ligand **6**.

Other structural evidence of the $Cs^+...\pi$ interactions has been reported by Thuery et al. [28] on the crystallographic study of the two dinuclear Cs complexes of the 1,3-calix[4]arene-*bis*-crown-6 (**5**) (Fig.7. 4) and of the 1,3-calix[4]arene-*bis*-benzo-crown-6 (**6**) (Fig. 7.5).

The environments of the cations are analogous in the two complexes. The Cs^+ ion interacts in a η^3 fashion with three terminal carbon atoms of the two inverted aromatic rings, with Cs-C distances ranging from 3.33(3) to 3.89(4) Å.

The cation...π interaction is also active when calixarene monoanions are used as ligands as shown by the crystallographic study of the monocaesium derivative of the bis(homooxa)-p-*tert*-butylcalix[4]arene (L), $[Cs(L-H)(OH_2)_3]\bullet 3H_2O$ [29], where the caesium cation is held in the calixarene cavity.

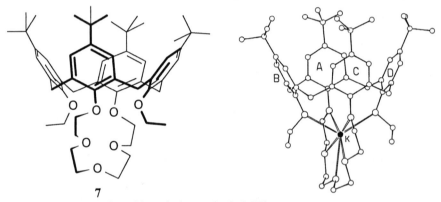

Figure 7.6. Perspective view of the cationic complex in **7**•KPic.

A crucial test which demonstrates how the cation...π interaction comes into play and its energy contribution to the complex stability has been obtained from the two X-ray crystal structures of the potassium picrate complexes of the 1,3-diethoxy-p-*tert*-butylcalix[4]arene-crown-5 (**7**) (see Fig. 7.6) in the *cone* conformation and that of the 1,3-diisopropoxy-p-*tert*-butylcalix[4]arene-crown-5 (**8**) in the *partial cone* conformation (see Fig. 7.7) [30]. It has been shown that in the *partial cone* complex there is a loss of one co-ordination site and that the K-O bond distances are significantly longer than those observed in the *cone* complex. Although one could lead to the conclusion that the *cone* conformer should bind the K^+ cation more strongly than the *partial cone* one.

Molecular mechanics calculations carried out on the *partial cone* complex have shown that the phenolic ring B, which is oriented towards the K^+ cation as in an ideal η^6 fashion, gives a contribution of *ca.* 6 kcal/mol to the stabilisation energy of

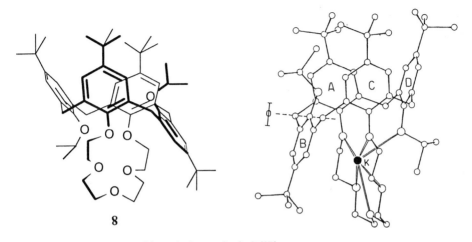

Figure 7.7. Perspective view of the cationic complex in 8•KPic.

the complex. It has also been demonstrated that in this case the deep minimum in the potential energy pattern is mainly due to the interaction between the electric field created by the cation and the electric dipole created by the electric field itself on the aromatic nucleus B faced on the cation [30]. This result has also shown that the cation... π interaction could be interpreted in a simpler way than the model of the "pseudo-anion" proposed by Dougherty [31] Our interpretation is also corroborated by a simple test on the data obtained by Dougherty itself by *ab initio* calculations on the alkali metal cations interacting with a benzene ring [31] which led to the equilibrium distances between the alkali metal cations and the benzene ring (r_o) and to the corresponding energy values (E). In fact, the products E r_o^2 give an almost constant value for the whole series of the alkali metal cations. This strongly indicates that the physical origin of the cation...aromatic interaction is of the charge...dipole type. Another question is why the *partial cone* ligands and not those in the *1,3 alternate* conformation prefer to bind the K^+ cation in this series of ligands derived from the p-*tert*-butylcalix[4]arenes. The *1,3 alternate* conformation, being less polar and less solvated, should maximise the cation...π interaction. In this case two *tert*-butyl groups facing the crown moiety in the *1,3 alternate* conformation reduce, for steric reasons, the stability of the complexes. For p-*tert*-butylcalix[4]arene derivatives, the *partial cone* structure seems to achieve the best compromise between stabilising cation...π interactions and destabilising steric effects due to the presence of alkyl groups in the *para* positions. The situation is quite different with p-Hcalix[arene derivatives. In fact, the 1,3-diisopropoxycalix[4]arene-crown-5 fixed in *the 1,3 alternate* conformation is a stronger K^+ binder and shows a K^+/Na^+ selectivity of ca. $3.4 \cdot 10^5$ better than valinomycin [32].

The cation...π interactions in which the π electrons are those of the internal walls of the calixarene cavity are also responsible for the complexation of the sodium cation by calix[4]arene derivatives. Transition metals such as Nb and Ta were used for shaping the calixarenes in the *cone* conformation and for electronically enriching the calixarene cavity [33]. The Na^+ cation is held by one η^6 and one η^3 interaction involving two opposite phenolic units almost parallel to each other within the calixene cavity.

7.3.1. π-arene complexes with transition metals.

Several other complexes in which a transition metal atom or ion interacts with one or more of the aromatic rings of the calixarene pocket are known. However, differently to that observed with alkali metal ions, where the ion interacts with the *endo* π surface of the calix, here the transition metal species interacts preferably with the *exo* π surface of the calix.

For example, hexacarbonylchromium $Cr(CO)_6$ reacts with all the four conformers of tetrapropoxycalix[4]arene (**9**) to give four different tricarbonyl-chromium complexes. The crystal and molecular structures of the *cone* **9**•$Cr(CO)_3$, *cone* **9**•$2Cr(CO)_3$ (which contains two crystallographically independent molecules) and *1,3 alternate* **9**•$Cr(CO)_3$, established by X-ray diffraction [34], are shown in Fig. 7.8.

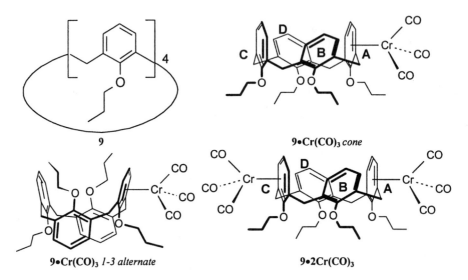

Figure 7.8. Structures of the tetrapropoxycalix[4]arene tricarbonylchromium complexes.

The main structural evidence, apart from the exo-π interaction mode of the metal, is the strong bending suffered by the chromium-bearing aryl unit A and its opposite unit C in the two *cone* complexes (the para carbons of the two rings are separated only by 3.81(1) Å in the *cone* 9•Cr(CO)$_3$ and 3.72(1) Å in the *cone* 9•2Cr(CO)$_3$). Consequently the two other rings, B and D, are strongly flattened and the calix[4]arene moiety takes the so called *bis-roof* structure which prevents any other inclusion within the intramolecular cavity. In the *1,3 alternate* 9•Cr(CO)$_3$ complex the *1,3 alternate* conformation of the calix appears not to be severely disturbed by the metal-arene interaction, so that in conclusion the severe distortion of the calix observed in the two *cone* structures seem to be mainly induced by the steric demand of the propyl groups, which interact with the neighbouring Cr(CO)$_3$ units.

That the *bis-roof* structure observed in the above cited chromium complexes is not induced by the exo π-complexation is also confirmed by the existence of dinuclear rhodium and iridium complexes of calix[4]arene derivatives in which the calix[4]arene is in the *cone* conformation and allows intramolecular complexation of organic neutral molecules [35].

Fig. 7.9 illustrates the dinuclear tricationic iridium complex in the *cone* conformation in which the two Ir-(η^5-C$_5$Me$_5$) groups interact with the exo π surfaces of two opposite aromatic rings without strong perturbations of the *cone* conformation so that each Et$_2$O guest molecule lies inside the hydrophobic cavity with its inner terminal Me group interacting *via* CH...π hydrogen bond with the two electron rich unmetallated aromatic rings. The CH...Ar(centroid) distances are: 3.53 3.69 Å (3.62 and 3.83 Å those with the metallated rings).

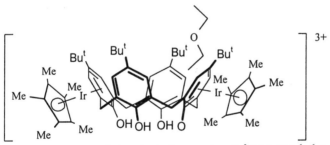

Figure 7.9. Molecular structure of the tricationic complex [(η^5-C$_5$Me$_5$)Ir(η^6:η^6-C$_{44}$H$_{55}$O$_4$)Ir(η^5-C$_5$Me$_5$)]$^{3+}$•Et$_2$O.

Two calix[4]arenetetrapropoxy (9) AgCF$_3$SO$_3$ complexes, in the *cone* and in the *partial cone* conformations, are the first examples of structurally characterised calix[4]arene-π complexes in which the metal interacts preferably with the endo π surface of the calixarene [36]-[37], as shown in Fig. 7.10.

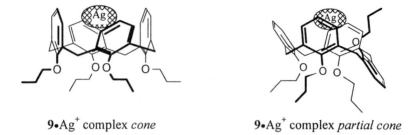

9•Ag$^+$ complex *cone*   9•Ag$^+$ complex *partial cone*

Figure 7.10. The structure of the two stereoisomers of calix[4]arenetetrapropoxy Ag$^+$ cationic complexes.

In the solid state, the cation complex in the *cone* conformation shows C_{2v} symmetry and the silver ion is held at the *upper rim* by the two para carbon atoms of the two almost parallel distal aryl rings. The Ag$^+$- *p*-C atom distances of the bound aryl rings are 2.39 and 2.40 Å and clearly reveal that the silver cation is held in the calix[4]arene cavity only through cation-π interactions. No bonding interaction with the metal cation has been observed with the other two aryl rings which are slightly but significantly flattened.

In the *partial cone* conformation the Ag$^+$ cation is again blocked at the *upper rim* between the two almost parallel distal aryl rings and the short Ag$^+$-*p*-C distances of 2.40 and 2.41 Å show that the cation interacts with the C=C$_{para}$ double bonds around the *p*-carbon atoms whereas the long Ag$^+$-*p*-C separation of 5.63 Å with the para carbon atoms of iso-oriented aryl ring rules out any π bonding with this ring. The short interatomic Ag-O distance of 2.94 Å between the Ag$^+$ ion and the phenolic oxygen of the inverted ring indicates that this ring also provides an electrostatic contribution to the stability of the complex. It is noteworthy that the *partial cone* conformation of the calix[4]arene in the complex is almost the same as that in the free ligand itself, thus indicating that this ligand is perfectly pre-organised for Ag$^+$ binding.

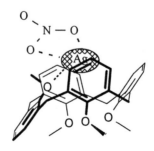

Figure 7.11. The molecular structure of the [tetrametoxycalix[4]arene Ag]$^+$ NO$_3^-$ complex.

Another complex of Ag^+ in the *partial cone* with the conformationally mobile tetrametoxycalix[4]arene has been structurally characterised (see Fig. 7.11) [38]. The silver ion lies at the *upper rim* of the calixarene held between the two distal aryl rings and interacts with the $C=C_{para}$ double bonds, but in contrast with the structure of the tetrapropoxy complex above described, the Ag^+ ion is further held by the phenolic oxygen of the inverted ring (Ag-O 2.517(3) Å).

The observed structures lead to the conclusion that in the *partial cone* structure the two distal arene rings are ideally pre-organised for Ag^+ binding. This is also confirmed by experiments that the conformationally mobile tetrametoxy calix[4]arene, which exists in solution in the *cone*↔*partial cone* equilibrium, undergoes a conformational change to the *partial cone* during the complexation of Ag^+.

7.4. Supramolecular Calixarene Complexes of Fullerenes.

Buckminsterfullerene and its related fullerenes are attracting increasing interest and have generated a new multidisciplinary research area [39]-[40]. In particular, one of the major targets is the design of bowl-shaped receptors for the selective complexation of the weak electron acceptor C_{60} molecule. Calixarenes have also proved to be excellent hosts for fullerenes. In particular, the selective formation of complexes between C_{60} and calix[8]arenes have been used for the purification of C_{60} and C_{70} [41]-[42]. ^{1}H NMR and IR measurements seem to suggest that the guest is stabilised by weak host-guest $CH_3...\pi$ interactions between the Bu^t of the host and the π surface of the guest.

Calix[5]arenes **10-12** strongly bind C_{60} in various organic solvents, always forming 1:1 host guest adducts [43], but in the solid state a contradictory host-guest ratio has been demonstrated by X-ray diffraction. In fact, the stoichiometry of the crystalline complex between **10** and C_{60} is 2:1 whereas those of the complexes between **11** and **12** with C_{60} are 1:1 [44].

10: X = I
11: X = Me
12: X = H

In the solid state the 2:1 complex $\mathbf{10_2 \cdot C_{60}}$ shows a highly symmetric pseudo D_{5d} cage structure (see Fig. 7.12) due to the statistic disorder of the two iodine atoms.

The **11•C$_{60}$** and **12•C$_{60}$** 1:1 complexes have roughly the same C_{5d} structure with the five-fold rotation axis of the host coincident with one of the five-fold rotation axes of the guest, and a five-membered ring of the guest is parallel to the mean plane of the five phenolic oxygens of the host. It is noteworthy that in all these crystalline complexes the free rotation of the C$_{60}$ guest is considerably suppressed although the atomic displacement of the atoms of the guest are significantly larger than those of the host. X- ray analyses on the three complexes indicate that there is no structural basis for the 2:1 complex formation of **10$_2$•C$_{60}$**.

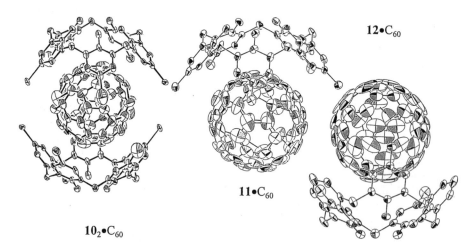

Figure 7.12. The molecular structures of the **10-12** complexes of C$_{60}$.

On the other hand, *ab initio* and molecular mechanics calculations carried out on the structures of the three complexes led to the HOMO energies (**10** –8.819, **11** – 8.843, **12** –8.875 eV) and to the van der Waals host-guest interaction energies (**10** – 3.51 , **11** –2.33, **12** –0.0 kcal mol^{-1}). A plot of the Gibbs energies of the complexation versus the van der Waals energies gave an excellent correlation in every solvent, thus suggesting that the van der Waals host-guest interactions are mainly responsible for the complexation.

More recently, the structure of the 2:1 complex suggested that two calix[5]arenes could be linked together through their *upper rim* to give shape selective receptors which prefer C$_{70}$ over C$_{60}$ [45].

Evaporation of toluene solutions of C$_{60}$ or C$_{70}$ in the presence of calix[6]arene (**1f**) yields crystals of [(calix[6]arene)(C$_{60}$)$_2$] or [(calix[6]arene)(C$_{70}$)$_2$] characterised by X-ray diffraction [46]. Both complexes crystallise in the space group $P4_12_12$ and are isostructural (see Fig. 7.13) despite the anisotropic shape of C$_{70}$. In each complex the calixarene is in the *double-cone* conformation and its associated shallow cavity is occupied by the fullerene guest disposed asymmetrically above its host cavities.

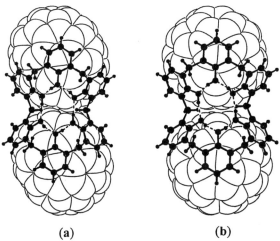

Figure 7.13. Projection of the structures of (a) [(calix[6]arene)(C_{60})$_2$] and (b) [(calix[6]arene)(C_{70})$_2$] both viewed from below the calixarene *double-cone*.[78]

The centre of the C_{60} guest molecule is as far as 7.06(7), 6.91(7) and 6.65(8) Å from the closest ring centroids. In the C_{70} complex the centre of the guest lies at 6.858(6), 7.23(7) and 7.46(7) Å from the closest ring centroids although its principal axis does not point toward the calixarene cavity. It must be emphasised that in both complexes the distances between calixarene and fullerene molecules are somewhat greater than the sum of the van der Waals radii. The molecular packings show that the calixarenes are surrounded by fullerenes on both sides with the calixarene nestled in a tetrahedral hole created by four neighbouring guest molecules.

Calix[4]arenes possess too small a cavity to form a ball and socket structure about a fullerene core, however, a co-crystallisation product between *p*-bromocalix[4]arene propyl ether and C_{60} has been reported. In the crystal lattice a remarkably well packed structure has been observed. It is formed by separate columns of C_{60} and *p*-bromocalix[4]arene propyl ether molecules with their dipole moments aligned unidirectionally [47]. This surprising packing mode allows the C_{60} molecules to undergo dipole induction, thus favouring the close packing of the fullerenes within their linear strands (the C_{60}...C_{60} separation is 0.05 Å shorter than in the structure of pure C_{60}).

The *p*-iodo calix[4]arene derivative **13** co-crystallises with C_{60} giving a multilayer structure formed by bilayers of calixarene intercalated with layers of C_{60} [48]. The calixarene molecule possesses C_2 symmetry and two distinct molecular orientations (rotated by 90° to each other) are present in each calixarene monolayer. The calixarene bilayer consists of two of the above described monolayers facing one another in a head-to-head fashion and leaving large intermolecular voids of sufficient size to be filled by the C_{60} molecules, as shown in Fig. 7.14.

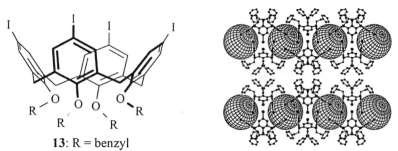

13: R = benzyl

Figure 7.14. Packing diagram viewed along [-110] showing the layers of C_{60} intercalated in the calixarene bilayers.[79]

It is intriguing that, although the long I...I contacts($\geq$ 4.0 Å, considerably longer i.e. than the value of 3.09 Å in the $C_{60}\bullet I_2\bullet$toluene) are comparable with the sum of the van der Waals radii, and that the closest calixarene C...C contacts are as long as 3.55 Å, the structure has produced a well ordered C_{60} molecule.

7.5. Crystal engineering.

7.5.1 Molecular networks based on molecular inclusion phenomena.

The prediction of the molecular packing of molecular solids is a fundamental step in the design of functional solids. Although at the present level of our knowledge, an overall understanding of the molecular interactions is far from satisfactory (thus preventing predictions of the packing mode in the crystalline phase [49]), simple molecular networks based on inclusion phenomena in the solid state can be designed. The fundamental idea behind this research is that, by analogy with molecules, (assemblies of atoms connected by covalent bonds), one can imagine a molecular network as one *hypermolecule* in which each component (a molecule or a complex formed by a molecule inside a molecule) is linked in the crystal lattice by non-covalent interactions. At the simplest stage the molecular network is designed in the step by step strategy based on the self-assembly of structurally pre-organised and complementary *tectons* [50]-[52].

7.5.2 Koilands and koilates.

Calix[4]arenes possessing a concave intramolecular cavity can be covalently linked together to give multiple linear cavity receptors termed *Koilands*. Linear Koilands can be assembled in infinite chains in the solid state using convex rigid connectors able to include their extremities in the intramolecular cavity of two neighbouring Koilands. Thus an infinite chain termed *Koilate* linked by non-covalent interactions is generated in the crystal lattice (see Fig 7.15).

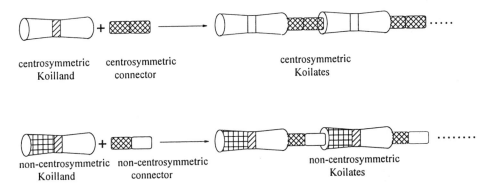

Figure 7.15. Schematic representation of the formation of Koilates using Koilands and connectors as building-blocks.

Examples of Koilands formed by two calix[4]arenes in the *cone* conformations covalently linked by silicon atoms [53] or by titanium(IV), niobium(V), aluminium(IV) or zinc [54]-[57] or also by organic [58]-[59] or by organometallic bridges, are known.

The molecular structures of the silicon-linked double p-methylcalix[4]arene **14** [60] and double p-*tert*-butylcalix[4]arene **15** [53] have been established by X-ray diffraction.

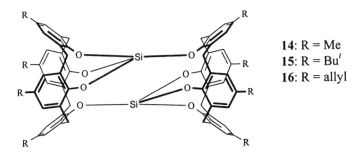

14: R = Me
15: R = Bu^t
16: R = allyl

X-ray study [61] showed that Koiland **15** forms linear koilates using, as linear connector, the rod type molecule of hexadine, which is 6.65 Å. long (distance from its two terminal Me groups). Koiland **16,** the silicon linked double pallylcalix[4]arene, does form van der Waals-assisted koilates with p-xylene as connectors. X-ray study [62] showed that the Koilates are linear and centrosymmetric with the two methyl groups of the connector deeply included inside the intramolecular cavity of two consecutive koilands (The shortest C-C distance between the guest methyl group and the nearest host carbon atom is 3.63 Å).

Hosseini *et al.* have shown that koilates can also be obtained by another strategy, using self-complementary Koilands bearing simultaneously two divergent and two connector moieties, as schematised in Fig. 7.16.

Figure 7.16. Schematic representation of an infinite molecular network based on the self-inclusion principle.

In the solid state, Koiland **16** does form an α-network with the inclusion of one of the four allyl groups at one face of the double calix[4]arene in the cavity of the

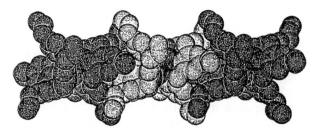

Figure 7.17. Perspective view of the Koilate from the self-inclusion of Koiland 16.[80]

consecutive one and held by van der Waals host-guest interactions [63] (see Fig 7.17). However, the Koilate formation is solvent-dependent; a change in the crystallisation solvent leads to the formation of discrete inclusion complexes.

7.5.3. Self-assembly assisted by $CH_3...\pi$ interactions.

The non-covalent attractive $CH_3...\pi$ interactions can be used for the self-assembly of monomeric units into columnar head-to-tail polymeric chains [64]. The two p-*tert*-butylcalix[4]arenes, titanium complexes **17** and **18**, form, in the solid state, columnar polymers in which each p-tolyl ligand of each complex is guested inside thecalix[4]arene cavity of a neighbouring complex below the titanium atom at 4.354(10) Å (**17**) and at 4.620(9) Å (**18**) so that the columnar axis coincides with the complex molecular axis as illustrated in Fig. 7.18.

The participation of attractive CH$_3$...π interactions between the methyl protons of the p-tolyl and two aromatic rings of the neighbouring complex molecule to the non-covalent head-to-tail interactions has been invoked. The C$_{p\text{-tolyl}}$...C$_{aromatics}$ distancesspread over a narrow range from 3.676(4) to 3.857(7) Å although they are weaker in the polymeric chains formed by the complex units **18**.

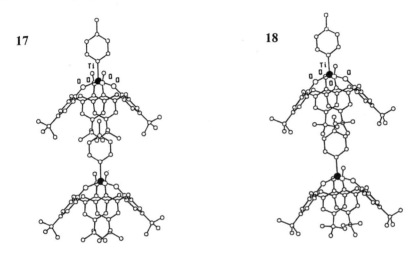

Figure 7.18. Perspective views of the complexes **17** and **18** along [010]. Primed atoms are obtained by the transformation -x,y,1/2-z.

Another example of self-assembled columnar head-to-tail polymeric chains formed by p-*tert*-butylcalix[4]arene wolframium complex (**19**) has been reported [65]. A wolframium atom bonded over the calix[4]arene oxomatrix has been functionalised with two phenoxo groups. When recrystallised from *n*-hexane (which is a non-competitive guest for the calix[4]arene cavity) one phenoxo group is guested within the calixarene cavity of an adjacent complex, simply faced on the host p-*tert*-butyl groups, thus leading to the columnar self-assembly reported in Fig. 7.19. However, the structural disorder affecting these groups made it impossible to establish whether the self-organisation is really assisted *via* CH$_3$...π interactions, although the shortest C...C host-guest contacts between the host *tert*-butyl carbons and the aromatic guest C atoms (3.49(2) and 3.62(2) Å) seem to support the participation of the attractive CH$_3$...π interactions.

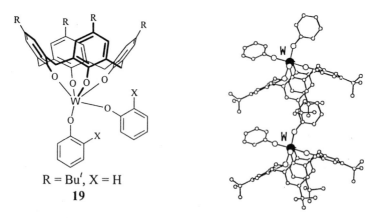

Figure 7.19. View of the polymeric head-to-tail chaining of complex **19** in the crystal lattice.

The CH$_3$...π interaction can be superimposed to electrostatic (much stronger) host-guest interactions. In this case the former plays a minor but not negligible rôle in determining the self-assembly in the solid state.

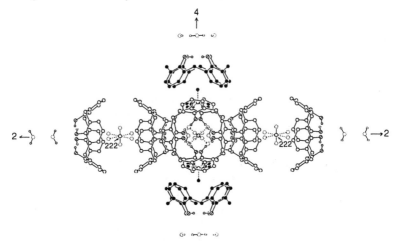

Figure 7.20. The stacking of the calixarene units A and B in the crystal lattice (the disordered [NMe$_4$]$^+$ ion is represented by dashed lines).[81]

An interesting example is the complexation of the tetramethyl ammonium ion with a calix[4]arene anion [66] whose crystal structure results from a subtle balance between electrostatic and CH$_3$...π interactions. In the solid state the compound crystallises as L•2[NMe$_4$]$^+$[L-H]$^-$•H$_2$O (L=calix[4]arene **1a**) and contains the [NMe$_4$]$^+$ cation (C) in two distinct environments, both involving inclusion within the calixarene cavity. One cation (C) is simply guested in a single calixarene (unit A) whereas the other is encapsulated by facing pairs of calixarenes (unit B). Units A and B are piled up in two different sequences intercalated by water molecules (W). The A moieties are stacked along a *4* axis in the sequence ...WACCAW..., while the B moieties are stacked along a *2* axis in the *ab* diagonal as the sequence ...WBCBW....(see Fig. 7.20).

7.5.4. Layered structures.

Layered structures formed by hydrophobic/hydrophilic interactions or self-assembled polymeric chains may be generated in the solid state by exploiting the original features of calix[4]arene diquinone **20** [67]. It is stabilised in the *1,3 alternate conformation* by the inclusion of a water molecule in the calixarene cavity (Fig. 7.21) and possesses one hydrophilic and one hydrophobic side. Moreover, two other water molecules are connected *via* hydrogen bonds to the nitrogen atoms at the hydrophilic end whereas no solvent molecules are at the hydrophobic side. In the crystal lattice the calixarene molecules and water solvent form a layered structure with alternatively hydrophobic/hydrophobic and hydrophilic/hydrophilic (aqueous layer) interactions.

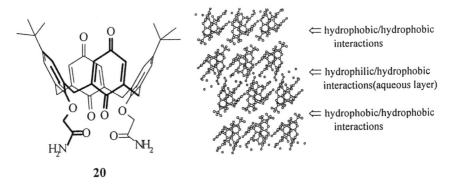

Figure 7.21. Packing diagram of the layered structure of the calix[4]arene diquinone in the *1,3 alternate* conformation **20**•3.5H$_2$O in the crystal lattice.

The calix[4]arene diquinone in the *cone* conformation forms a NaClO$_4$ complex with the calixarene [67]. In such conditions cation quinone oxygen atom coordination occurs at both the *upper* and the *lower rim*. The sodium atom, which is seven coordinate, is bonded to the six oxygen atoms at the *lower rim* (Na-O$_{quinone}$ 2.315(9), 2.306(6) Å, Na-O$_{carbonyl}$ 2.380((9), 2.315(7) Å, Na-O$_{ether}$ 2.367(7) 2.537(6) Å) and to one *upper-rim* quinone oxygen from an adjacent molecule at 2.410(7) Å. Thus the complex forms in the crystal lattice the one-dimensional chain illustrated in Fig.7. 22.

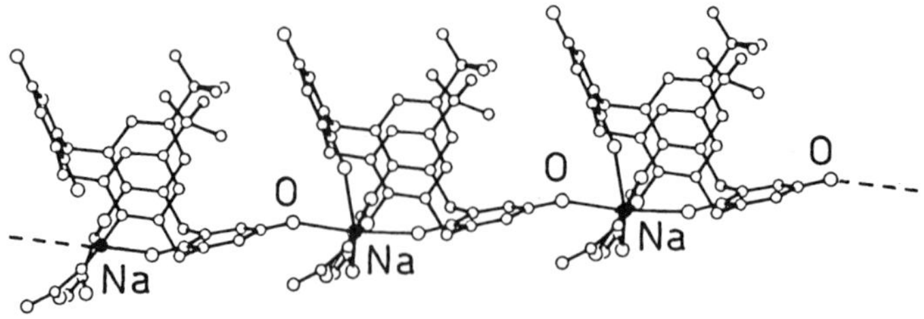

164

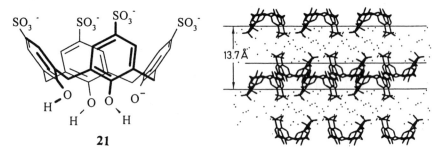

Figure 7.22. Crystal structure of the structure of calix[4]arene diquinone(in the *cone* conformation)•NaClO$_4$ complex showing the one-dimensional coordination.

Water soluble calix[4]arene derivatives or complexes self-organise in the solid state in multilayer structures with intercalation of cationic, anionic and molecular species between calixarene layers. The sodium salt of the calix[4]arene sulfonate **21** pentaanion exists in the solid state in the highly ordered multilayer structure **21**•5Na$^+$•12H$_2$O, which resembles a bio-organic bilayer structure [68] (see Fig. 7.23).

Figure 7.23. Side view of the multilayer structure of **21**•5Na$^+$•12H$_2$O in the lattice.[82]

The structure consists of organic layers each formed by up-down calixarenes, a sodium ion and a water molecule simultaneously interacting with the negatively charged groups of the calixarenes, whereas the inorganic layer consists of the remaining four 5Na$^+$ cations and eleven water molecules all organised through a complicated network of hydrogen bonds. A similar self-assembly can be observed if one recrystallises the sodium calix[4]arene sulphonate from a water/acetone solution. The **21**•5Na$^+$•8H$_2$O•Me$_2$CO product similarly grows through layers of calixarene anions, guesting within their cavities the acetone molecules, intercalated with layers of sodium ions and water molecules. The thickness of the hydrated layers are 8.3 and 8.4 Å in the calixarene/water and in the calixarene/water/acetone structure respectively. It is noteworthy that the sodium ions are exchangeable, the structures of the K$^+$, Rb$^+$, Cs$^+$ parents being similar. The ammonium ion is also exchangeable, although the structure of its complex is slightly different because of the displacement of water by the ammonium cations [69]. In fact, the polar layer is no longer formed of four alkali metal cations and eleven water molecules but is composed of five ammonium ions, one water molecule and a methyl sulphate anion, with the methoxy group of the latter within the calixarene cavity held by hydrogen bonds.

The lattices of three transition metal complexes (Cr, Yb and Cu) of water-soluble calixarene (**21**) showed layered structures although the incorporation modes of the metals in the gross structure are completely different from each other [70]. In the crystal lattice of the layered Cr(III) complex $[Cr(OH_2)_6][Na][calix[4]arene sulphonate](acetone) \bullet 10.5H_2O$, the acetone molecule is guested within the calixarene cavity and the $Cr(OH_2)_6^{3+}$ ions are intercalated within the hydrophilic layer. In the case of the ytterbium(III), $[Yb(OH_2)_7][Na][calix[4]arene sulphonate](acetone) \bullet 9H_2O$ structure, each $Yb(OH_2)_7^{3+}$ ion is directly bonded to a sulphonate oxygen. An interesting feature of this structure is the architectural rôle of one water molecule included within the calixarene cavity. This water molecule weakly links the hydrophilic to the hydrophobic layer, being hydrogen bonded (at 2.76 Å) to a water molecule of the first co-ordination sphere of the ytterbium and to another water molecule in the hydrophilic layer and at the same time weakly π-bonded to the aromatic walls of the calixarene cavity (3.48, 3.70, 3.70, 3.80 Å from the aromatic centroids).

Quite different is the incorporation mode of the transition metal in the multilayer solid state structure of the $\{[Yb(OH_2)_4][Na]_2[calix[4]arene-sulphonate](acetone) \bullet 7.5\ H_2O\}_2$. Two different complex units, with two different copper environments, are present in the structure of the $\{[Cu(OH_2)_4][Na]_2[calix[4]arene sulphonate] \bullet 7.5\ H_2O\}_2$ complex where the $Cu(OH_2)_4^{2+}$ ions units interact with the sulphonate oxygens of different calixarene molecules in both an *intra-* and *inter*-layer fashion.

7.5.5. Self-assembly via self-inclusion assisted by CH...π intermolecular interactions.

As described above, another important, although not very flexible, tool in crystal engineering is the self-inclusion between adjacent molecules in a crystal lattice.

As an example of this approach we here report the crystal structure of p-*tert*-butylcalix[5]arene **1e** [71] in which the calix[5]arene cavity guests, *via* weak CH...π arene interactions, the *tert*-butyl group of a neighbouring calixarene related by a *c*-glide plane symmetry, thus generating, in the lattice, a zigzag molecular chain along the *c*-glide plane, as shown in Fig. 7.24.

Self-assembled polymeric chains are also generated via self-inclusion between adjacent molecules of the p-*tert*-butylcalix[4]arene bis-ethyl ester **23** [72] (see Fig. 7.25).

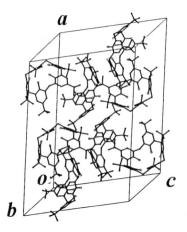

Figure 7.24. Molecular chains in the crystal lattice of p-*tert*-butylcalix[5]arene **1e** as viewed along their axes.

The self-inclusion process of the ester residues in the cavity of the adjacent molecule is supported by weak CH...π arene hydrogen-bond interactions between the hydrogen atoms of the terminal methyl group of the ester moiety of one calixarene and the π arene electrons of one adjacent calixarene cavity.

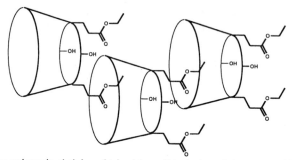

Figure 7.25. The polymeric chaining obtained by self-inclusion of the ester residue in one adjacent cavity of calixarene **23**.

The intermolecular weak CH...π arene hydrogen-bond interactions have also been invoked to explain the polymeric self-assembly determined by the self-inclusion of the phthaloyl residue of the calix[6]arene **24** in one adjacent calix[6]arene molecule [73]. However, here the CH...π arene interactions are double: a) the aromatic hydrogen atoms of the phthaloyl group interacts with the aromatic π clouds of the adjacent calix[6]arene basket and b) the hydrogen atoms of the methyl groups of the adjacent calix[6]arene interacts with the π electrons of the phthaloyl moiety.

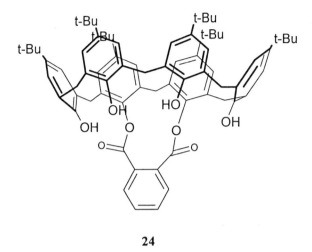

24

7.6. Second-sphere co-ordination.

According to Stoddart [74] the second-sphere co-ordination may be defined as *"the non-covalent bonding of chemical entities to the first co-ordination sphere of a transition metal complex"*. Thus the second co-ordination sphere concept with calixarenes can be used to build up, for example, bilayer structures. A host-guest complex which involves second-sphere co-ordination has been reported by Atwood and Hamada [75] for the structure of $[(H_2O)_4Cu(NC_5H_5)_2](H_3O)_3[calix[4]arene$ sulphonate]•$10H_2O$ (see Fig. 7.26) in which the primary co-ordination sphere of the Cu^{2+} ion is composed of two pyridine and four aqua ligands.

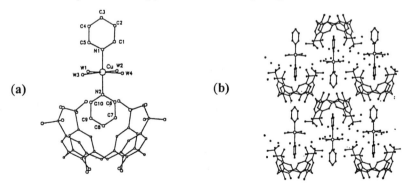

Figure 7.26. (a) $[(H_2O)_4Cu(NC_5H_5)_2]^{2+}$ as a guest in [p-sulphonatocalix[4]arene]$^{5-}$ (b) bilayer structure in the crystal packing of $[(H_2O)_4Cu(NC_5H_5)_2](H_3O)_3$- [calix[4]arene sulphonate]•$10H_2O$.[83]

The second co-ordination is achieved, *via* van der Waals interactions, by one pyridine ring guested within the calixarene cavity, while the other is intercalated into the bilayer. Hydrophilic layers of copper complexes are intercalated to layers of calixarenes. Another example of bilayer self-assembly based on the second-co-ordination sphere strategy is the structure of $[(H_2O)_5Ni(NC_5H_5)]_2(Na)[calix[4]arene$ sulphonate]•3.5 H_2O [76] in which the hydrophobic cavity of the [calix[4]arene sulphonate]$^{5-}$ is used for the second-sphere co-ordination while the Ni complex spans the hydrophilic layer.

7.7. Supramolecular assemblies organised by feeble forces.

The structure of the Na_4(pyridinium)[calix[4]arenesulphonate]•8H_2O complex [77], reported in Fig. 7.27, is another example of self-assembly based on weak interactions. Two (pyridinium)[calix[4]arenesulphonate complexes are linked in a dimer via inter-complex N-H...O hydrogen bonds

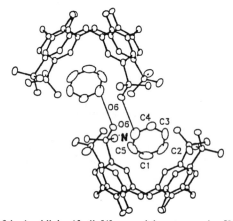

Figure 7.27. Structure of the (pyridinium)[calix[4]arenesulphonate complex.[84]

The structure of the $[Cu(NC_5H_5)]_2(H_2O)_3(Na)_3[calix[4]arene$ sulphonate]•13H_2O inclusion complex [77], shown in Fig. 7.28, is formed by super-dimers in which the copper(II) ion is strongly co-ordinated to two pyridine nitrogens (1.98(1), 2.00(1) Å) and two water oxygen atoms (2.00(1), 2.01(1) Å).

The remaining two axial co-ordination sites of the strongly distorted octahedral arrangement are occupied by one water molecule (Cu...O 2.50(1) Å) and by a sulphonate oxygen atom (2.55(1) Å. The latter weak interaction binds the copper complex to the exterior of the calixarene, thus one of the two pyridine ligands fills the cavity of an adjacent calix[4]arene. The super-dimer architecture is the result of the simultaneous co-operation of two weak Cu...sulphonate bonds and two hydrophobic interactions.

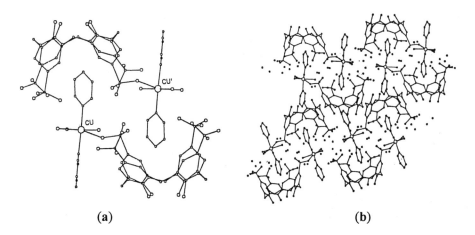

Figure 7.28. Structure of the $[Cu(NC_5H_5)]_2(H_2O)_3(Na)_3[calix[4]arene-sulphonate] \cdot 13H_2O$ inclusion complex. (a) The super-dimer formation *via* a secondary Cu...O sulphonate bond. (b) Packing diagram of the complex. The Na^+ ions and water molecules in the interlayer space are represented by full circles.[84]

7.8. References.

[1] Andreetti G.D., Ungaro R., Pochini A. *J. Chem. Soc. Chem. Commun.* (1979) 1005-1006.
[2] Facey G.A., Dubois R.H., Zakrewski M., Ratcliffe C. I., Atwood J.L., Ripmeester J. A. *Supramol. Chem.* **1** (1993) 199-200.
[3] Andreeetti G.D., Ugozzoli F. unpublished result.
[4] Brower E.B., EnrigthG.D., Ratcliffe C.I., Ripmeester J.A. *Supramol. Chem.* **7** (1996) 79-83.
[5] Perrin M., Gharnati F., Oehler D., Perrin R., Lecocq S.*J. Inclusion Phenom. Mol. Recognit. Chem.* **14** (1992) 257-270.
[6] Ohtsuchi M. *et al. Acta Crystallogr.* **C49** (1993) 639-641.
[7] Ungaro R., Pochini A., Andreetti G.D., Domiano P., J. Chem. Soc. Perkin Trans. 2 (1983) 1773-1779.
[8] Ungaro R., Arduini A., Casnati A., Ori O. Pochini A. Ugozzoli F. in *Computational Approaches in Supramolecular Chemistry*, Wipff G. Ed., NATO ASI Serie C Vol 426, (Kluwer Academic Publishers, The Netherlands, 1993) 277-300, and references therein.
[9] Andreetti G.D., Ori O., Ugozzoli F., Alfieri C., Pochini A., Ungaro R. *J. Inclusion Phenom.* **6** (1988) 523-536.
[10] Kollman P.A., Mc Kevery J., Johansson A., Rothemberg S. *J. Am. Chem. Soc.* **97** (1973) 955-965.
[11] Press W., *Single Particle Rotations in Molecular Crystals* (Springer, Berlin, 1981).
[12] (a) Caciuffo R. *et al.*, *Phys. B* **180& 181** (1992), 691-693. (b) Caciuffo R. *et al.*, *Chem. Phys Lett.* **201-5,6** (1993), 427-432. (c) Paci B. *et al. Phys Chem. A* **102** (1998) 6910-6915.
[13] Caciuffo R. *et al. Mol. Phys.* **81-3** (1994) 609-619.
[14] Caciuffo R. *et al. Phys. B* **202** (1994) 279-286.
[15] Caciuffo R. *et al. Phys. B* **234-236** (1997) 115-120.

[16] Schneider H.-J., Blatter T., Zimmerman P. *Angew. Chem. Int. Ed. Engl.* **29** (1990) 1161-1162.
[17] Petti M.A., Shepodd T.J., Barrans R. E. Jr., Dougherty D.A., *J. Am. Chem. Soc.***110** (1988) 6825-6840.
[18] Stauffer D.A., Barrans R.E. Jr., Dougherty D.A., *Angew. Chem., Int. Ed. Engl.* **29** (1990) 915-918.
[19] Mc Curdy A., Jimenez L., Stauffer D.A., Dougherty D.A., *J. Am. Chem. Soc.***114** (1992) 10314-10321.
[20] Ikeda A., Shinkai S., *J. Am. Chem. Soc.***116** (1994), 3102-3110 and references therein.
[21] Koh K. N., Araki K., Ikeda A., Otsuka H., Shinkai S. *J. Am. Chem. Soc.***118** (1996) 755-758.
[22] Sussman J.L. *et al. Science* **253** (1991) 872-879.
[23] For a recent review see: Dougherty D.A. *Chem. Rev.* **97** (1997) 1303-1324.
[24] Ungaro R. *et al. Angew. Chem. Int. Ed. Engl.* **33** (1994) 1506-1509.
[25] Casnati A. *et al. J. Am. Chem. Soc.***117** (1995) 2767-2777.
[26] Gregory K. *et al. Angew. Chem. Int. Ed. Engl.* **28** (1989) 1224-1226.
[27] Harrowfield J.M. *et al. J. Chem. Soc. Chem. Commun.* (1991) 1159-1161.
[28] Thuery P. *et al. J. Incl. Phen. Mol. Rec. Chem.* **23** (1996) 305-312.
[29] Assmus R. *et al. J. Chem. Soc. Dalton Trans.* (1993) 2427-2433.
[30] Ugozzoli F. *et al. Supramol. Chem.* **5** (1995) 179-184.
[31] Kumpf A., Dougherty D.A. *Science* **261** (1993) 1708-1713.
[32] Casnati A. *et al. Chem. Eur. J.* **2** (1996) 436-445.
[33] Zanotti A. et al. *J. Chem. Soc. Chem. Commun.* (1997) 183-184.
[34] Iki H., Kikuchi T., Tsuzuki H. Shinkai S. *J. Incl. Phen. Mol. Rec. Chem.* **19** (1994) 227-236.
[35] Steed J.W., Juneja R.K., Burkhalter R.S., Atwood J.L. . *J. Chem. Soc. Chem. Commun.* (1994) 2205-2206.
[36] Ikeda A., Tsuzuki H., Shinkai S. *J. Chem. Soc. Perkin Trans.* 2 (1994) 2073-2080.
[37] Ikeda A. and Shinkai S. *J. Am. Chem Soc.* **116** (1994) 3102-3110.
[38] Xu W., Puddephatt R.J., Muir K.W., Torabi A.A. *Organometallics* **13** (1994) 3054-3062.
[39] Krätschmer W, Lamb L.D., Fostiropoulos K., Huffman D.R.. *Nature* **347** (1990) 354-358.
[40] Kroto H.W., *Angew. Chem. Int. Ed. Engl.* **104** (1992) 113-133.; *Angew. Chem. Int. Ed. Engl.* **31** (1992) 111-129.
[41] Suzuki T., Nakashima K., Shinkai S. *Chem. Lett.* (1994) 699-702.
[42] Atwood J.L., Koutsantonis G.A., Raston C.L. *Nature* **368** (1994) 229-231.
[43] Haino T., Yanase M., Fukazawa Y. *Angew. Chem. Int. Ed. Engl.* **36** (1997) 259-260.
[44] Haino T., Yanase M., Fukazawa Y. *Tetrahedron Lett.* **38** (1997) 3739-3743.
[45] Haino T., Yanase M., Fukazawa Y. *Angew. Chem. Int. Ed. Engl.* **37** (1998) 997-998.
[46] Atwood J.L., Barbour L.J., Raston C.L., Sudria I.B.N., *Angew. Chem. Int. Ed. Engl.* **37** (1998) 981-983.
[47] Barbour L.J., Orr G.W., Atwood J.L. *Chem Commun.* (1998) 1901-1902.
[48] Barbour L.J., Orr G.W., Atwood J.L. *Chem Commun.* (1997) 1439-1440.
[49] Gavezzotti A. *Acc. Chem. Res.* **27** (1994) 309-314.
[50] Mann S. *Nature* **365** (1993) 499-505.
[51] Philip D., Stoddart J.F. *Angew. Chem. Int. Ed. Engl.* **35** (1996) 1155-1196.
[52] Simard M., Su D., Wuest J.D. *J. Am. Chem. Soc.* **113** (1991) 4696-4698.
[53] Delaigue X. *et al. Tetrahedron Lett.* **34** (1993) 3285-3290.
[54] Olmstead M.M., Sigel G., Hope H., Xu X., Power P. *J. Am.Chem.Soc.* **107** (1985) 8087-8091.
[55] Corazza F., Floriani C., Chiesi-Villa A., Guastini C., *J. Chem. Soc. Chem. Commun.* (1990) 1083-1084.
[56] twood J.L., Bott. S.G., Jones C., Raston C.L. *J. Chem. Soc. Chem. Commun.* (1992) 1349-1351.
[57] twood J.L., Junk P.C., Lawrence S.M., Raston C.L. *Supramol Chem.* **7** (1996) 15-17.
[58] raft D. *et al. Tetrahedron lett.* **31** (1990) 4941-4944.
[59] an Loon J.-D. *et al. J. Org. Chem.* **55** (1990) 5176-5179.
[60] Hajek F., Graf E., Hosseini M.W. *Tetrahedron Lett.* **37** (1996) 1409-1412.

[61] Hajek F., Graf E., Hosseini M.W. *Tetrahedron Lett.* **37** (1996) 1401-1404.
[62] Hajek F., Graf E., Hosseini M.W., De Cian A., Fischer J. *Angew. Chem. Int. Ed. Engl.* **36** (1997) 1760-1762.
[63] Hosseini M.W., De Cian A.*Chem. Commun.* (1998) 727-733.
[64] Zanotti-Gerosa A. *et al. Inorg. Chim. Acta* **270** (1998) 298-311.
[65] Zanotti-Gerosa A. *et al. J. Chem Soc. Chem Commun.* (1996) 119-120.
[66] Harrowfield J.M., Richmond W.R., Sobolev A.N., White A.H. *J. Chem Soc. Perkin Trans 2* (1994) 5-9.
[67] Beer P.D. *et al. Inorg. Chem.* **36** (1997) 5880-5893.
[68] Coleman A.W. *et al. Angew. Chem. Int. Ed. Engl.* **27** (1988) 1361-1362.
[69] Bott S., Coleman A.W., Atwood J.L. *J. Am.Chem.Soc.* **110** (1988) 610-611.
[70] Atwood J.L. *et al. Inorg. Chem.* **31** (1992) 603-606.
[71] Gallagher J.F., Ferguson G., Böhmer V., Kraft D, *Acta Cryst* **C50** (1994) 73-77.
[72] Ferguson G. *et al. Supramol. Chem.* **7** (1996) 223-228.
[73] Kraft D., Böhmer V., Vogt W., Ferguson G., Gallagher J.F. *J. Chem. Soc. Perkin Trans 1* (1994) 1221-1230.
[74] Alston D.R., Slawin A.M.Z., Stoddart J.F., Williams D.J. *Angew. Chem. Int. Ed. Engl.* **24** (1985) 786-787.
[75] Atwood J.L *et al. Pure & appl. Chem.* **65** (1993) 1471-1476.
[76] Atwood J.L. *et al. J. Am. Chem. Soc.* **113** (1991) 2760-2761.
[77] Atwood J.L., Orr G. W., Hamada F., Bott S.G., Robinson K.D. *Supramol. Chem.* **1** (1992) 15-17.
[78] Reprinted from ref. 46 with permission of Wiley-VCH.
[79] Reprinted from ref. 48 with permission pf the Royal Society of Chemistry.
[80] Reprinted from ref. 63 with permission pf the Royal Society of Chemistry.
[81] Reprinted from ref. 66 with permission pf the Royal Society of Chemistry.
[82] Reprinted from ref. 75 with permission of International Union of Pure and Applied Chemistry.
[83] Reprinted from ref. 77 with permission of Gordon and Breach Science Publishers.

CHAPTER 8

CALIXARENES IN THIN FILM SUPRAMOLECULAR MATERIALS

ANDREW J. LUCKE AND CHARLES J. M. STIRLING

*Department of Chemistry, University of Sheffield, Brook Hill, Sheffield S3 7HF
United Kingdom*

Calix[n]arenes have found new applications in the area of thin films. Thin films can be thought of as two dimensional supramolecular materials that are of both fundamental and industrial interest. Calix[n]arenes and resorcin[4]arenes can be deposited as thin films through a variety of techniques such as Langmiur-Blodgett, self-assembly, and evaporative methods. This chapter examines the physical properties of thin films and their interactions with their environment. In particular, thin films of calix[n]arenes and their adsorption of metal ions, molecular ions, and neutral organic molecules have become of great interest for use in new devices for sensing, nonlinear optics, pyroelectric and semiconductor applications.

8.1 Introduction

Calix[n]arenes and resorcin[4]arenes (Figure 1) have become important molecules for the preparation of new supramolecular materials.[1-4] These novel materials have been examined for purely fundamental interest but, often they have other potential applications. Central to their ongoing development as new supramolecular materials are their host-guest properties. Calix[n]arenes and resorcin[4]arenes have found applications in thin films, chromatography supports, polymers and catalysts. This review focuses on the preparation and application of thin films of calix[n]arenes and resorcin[4]arenes.

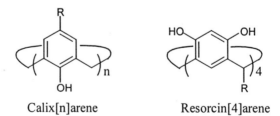

Calix[n]arene   Resorcin[4]arene

Figure 1. Generalised structures of calix[n]arenes and resorcin[4]arenes.

8.2 Thin Films of Calix[n]arenes

Calix[n]arenes have been used as building blocks for new supramolecular materials. Often these new materials take the form of thin films, coatings and supported materials. Many of the new supramolecular materials have been designed and examined for their binding abilities for use in sensors and as complexing agents. In a solid state type of commercial device these calix[n]arenes would generally be used as a thin film that interacts with its immediate environment in a measurable way. Calix[n]arenes have been used to form Langmuir-Blodgett (LB) mono- and multilayers, as well as self-assembled mono- and multilayers. They are also suitable for a variety of evaporative techniques. Therefore, the physical and binding properties of these new materials required investigation.

Amphiphilic calix[n]arenes and resorcin[4]arenes, i.e. bearing both hydrophobic (generally long alkyl chains) and hydrophilic groups (Figure 2) are generally suitable for the formation of LB films. The formation of a Langmuir film involves the spreading of a calix[n]arene in a volatile solvent onto the surface of an aqueous subphase. Following evaporation of the solvent, a disordered monolayer exists over the surface of the water which can be compressed to form stable ordered monolayers on the water surface. These Langmuir films can then be transferred to suitable surfaces to form LB mono- or mulitlayers.

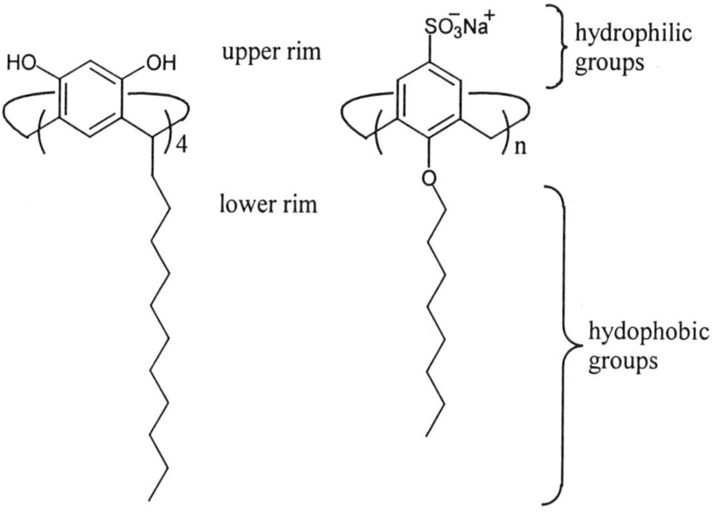

Figure 2. Examples of an amphiphilic resorcin[4]arene and calix[n]arene.

The first reports of calix[n]arene monolayer films began to appear during the late 1980's and since then a variety of calix[n]arenes have been used to form LB films. Regen et al. have reported the formation of Langmuir monolayers at the air-water interface from mercurated O-alkylated calix[6]arenes derived surfactants **1-6**.[5,6]

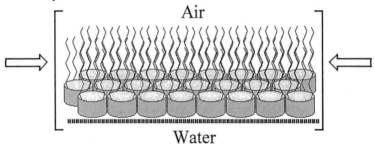

1 X = HgO_2CCF_3,   R = C_2H_5
2 X = HgO_2CCF_3,   R = n-C_4H_9
3 X = HgO_2CCF_3,   R = n-C_8H_{17}
4 X = HgO_2CCF_3,   R = n-$C_{16}H_{33}$
5 X = C(NOH)NH_2,   R = n-C_8H_{17}
6 X = CONHCH$_2$CH$_2$SSCH$_3$,
       R = n-C_8H_{17}

These calix[n]arenes formed stable compressed monolayers with properties consistent with the formation of a hexagonally packed array of molecules. The area per molecule was of the order of 150 ±8 Å^2 in agreement with that predicted by CPK models. The films were shown to be porous in character through water evaporative experiments.[5]

Figure 3. Illustration of a compressed Langmuir monolayer of calix[6]arene **3** at the air water interface.

The films could be made more cohesive through the crosslinking of the calix[n]arenes by the addition of malonic acid to the subphase.[5,6] The length of the O-alkyl chain was shown to affect the compressibility of the calix[n]arenes, small ethyl substituents **1** produced a less condensed layer than n-octyl **3** or n-hexadecyl **4** calix[n]arenes which produced highly condensed monolayers. This indicated that strong intermolecular hydrophobic interactions were playing an important role in defining the packing density of the film.[6] Langmuir monolayers could also be prepared from similar calix[n]arenes where n = 4, 5, and 7.[6] Further work by Regan et al. showed that calix[6]arene **6**[7,8] and in particular a hydroxylamine derived calix[6]arene **5**[9] also formed stable monolayers and that LB multilayers of these calix[n]arenes on suitable polymer supports could be used for permeation-selective (e.g. He/SF$_6$ selectivity >440) supramolecular composite materials.[9] Recently a boronic acid-based calix[6]arene was also reported as having an extraordinary cohesiveness at the air-water interface through intermolecular H-bonding and/or acid-base interactions.[10]

Shinkai et al. showed that Langmuir monolayers could be formed by *p*-octadecyl-calix[4]arenes and ester functionalised calix[n]arenes.[11]

Coleman et al. examined the stability of Langmuir monolayers of calix[4]arenes **7-10** as a function of substitution at the upper (through *p*-C_{11} acyl substitution) and lower rim (through addition of ethyl bromoacetate).[12] They found their monolayer stability increased in the substitution order **7 < 8 ≈ 9 << 10**.

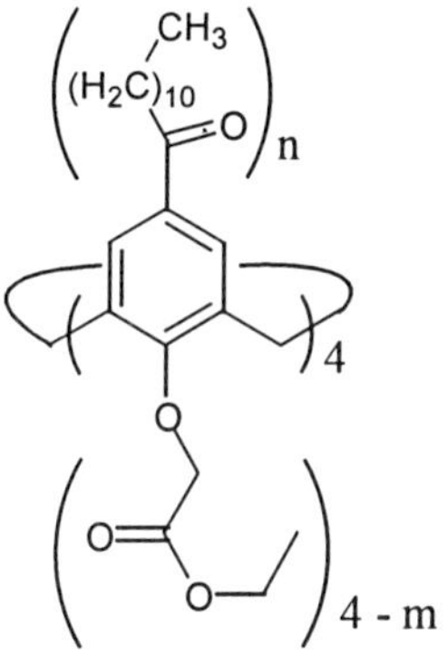

7	n = 1, m = 3
8	n = 1, m = 0
9	n = 4, m = 3
10	n = 4, m = 0

(m = number of unsubstituted hydroxy groups)

The calix[4]arene **7** substituted at either end was thought to orientate itself parallel to the air-water interface (Figure 4) with the alkyl substituents perpendicular to the surface.[12]

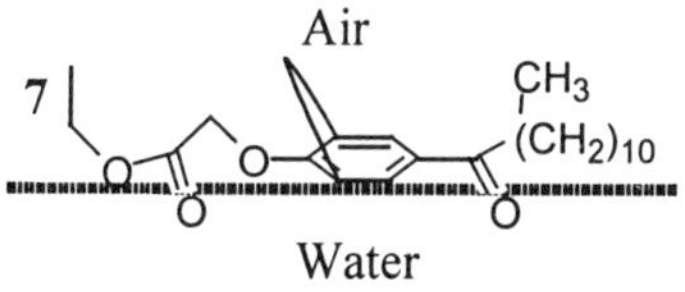

Figure 4. Schematic representation of calix[4]arene **7** monolayer on water.

A variety of resorcin[4]arenes have also been shown to form LB mono- and multilayers.[13-19] Dutton et al. have shown that resorcin[4]arenes **11** and **12** form stable LB films but that their spectroscopic behavior depended on the spreading solvent.[14] In addition, the authors reported that if metal cations (e.g. Hg^+, Cu^+, Cd^+, Na^+ and K^+) were present in the aqueous subphase changes in the resorcin[4]arene conformation occured.

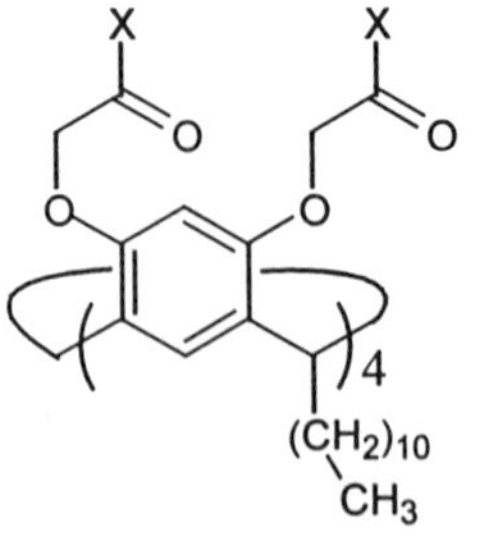

11 X = OCH$_3$
12 X = NEt$_2$

Stirling et al. have studied LB mono- and multilayers of a variety of resorcin[4]arenes **13-25** and cavitands **27** and **28**.[18] The authors were able to show that a wide variety of structure in both the rim of the bowl and in the pendant legs could be tolerated in the formation of Langmuir monolayers.

Resorcin[4]arene	Number	R	X	Y
	13	-C$_{11}$H$_{23}$	H	OH
	14	-(CH$_2$)$_8$CH=CH$_2$	H	OH
	15	-C$_{17}$H$_{35}$	H	OH
	16	-(CH$_2$)S(CH$_2$)$_{10}$H	H	OH
	17	-(CH$_2$)$_8$CH=CH$_2$	OH	OH
	18	-(CH$_2$)$_8$CH=CH$_2$	CH$_3$	OH
	19	-C$_{11}$H$_{23}$	OH	OH
	20	-C$_{11}$H$_{23}$	HgOAc	OH
	21	-(CH$_2$)$_8$CH=CH$_2$	H	OSiMe$_3$
	22	-(CH$_2$)$_8$CH=CH$_2$	H	OCOCH$_3$
	23	-(CH$_2$)S(CH$_2$)$_{10}$H	H	OCOCH$_3$
	24	-C$_6$H$_5$	H	OCOCH$_3$
	25	-(CH$_2$)$_{10}$OH	H	OH
	26	-(CH$_2$)$_{10}$SH	H	OH

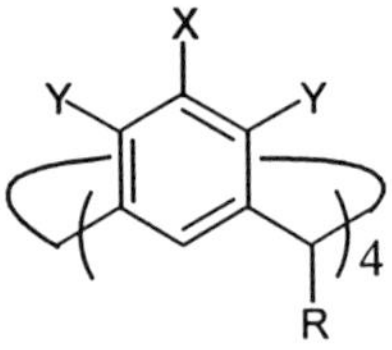

Furthermore, the majority of these derivatives could be deposited onto substrates as multilayers. Derivatisation at the resorcin[4]arene phenolic positions by groups like trimethylsilyl or acetyl groups (e.g. **21** and **22** respectively) notably reduced the monolayer collapse pressures (i.e. the pressure at which the ordered monolayers lift from the surface and collapse) and increased the observed area per molecule. This indicates that in phenolic-functionalised resorcin[4]arenes the bowl like cavity was destabilized from a cone to a flattened cone conformation on the surface.[18]

Cavitand	Number	R	Y
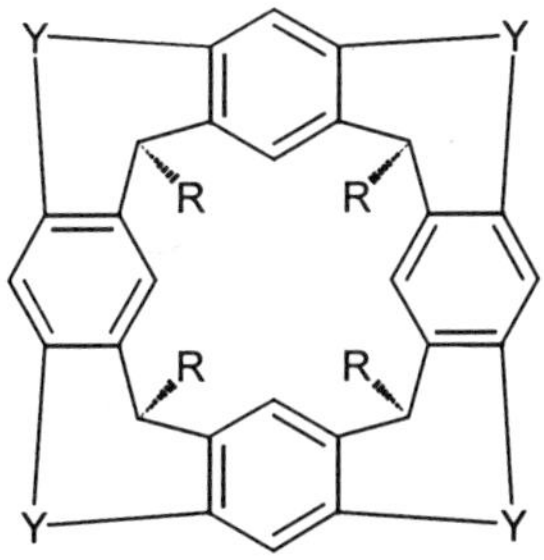 (structure)	27	-(CH$_2$)$_8$CH=CH$_2$	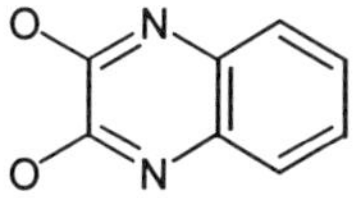 quinoxaline-dioxy group
	28	-(CH$_2$)$_8$CH=CH$_2$	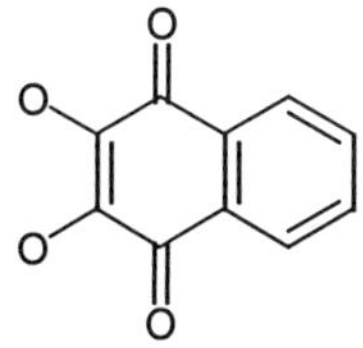 naphthoquinone-dioxy group

The cavitand series of compounds **27** and **28** had rigid structures that displayed steep Langmuir isotherms indicating a well packed monolayer. This rigidity however produced poorly formed LB multilayers.[18]

The most interesting feature of this work arises from resorcin[4]arene **25** that contains both polar phenolic hydroxy groups at the head group and long hydroxy terminated alkyl chain legs. This compound showed novel behavior in regions of the π-A isotherm when compressed on the water subphase. LB mltilayers could be deposited at both high and low pressures. Contact angle, grazing angle FTIR, and x-ray reflectometry experiments suggested that at low pressures, the hyroxylated legs curl over to come into contact with the water surface while at high pressures the compounds may be forced into the structures as shown. Area per molecule measurements appear to fit model (**b**) in an alternating configuration.[18]

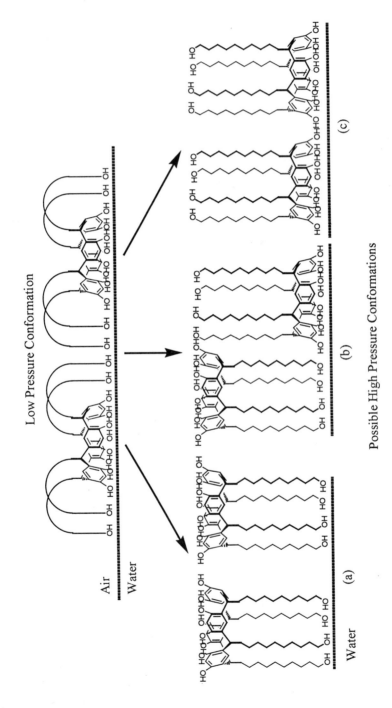

Figure 5. Possible low- and high-pressure (a-c) arrangements of resorcin[4]arene **25** on the water surface.

Ichimura et al. have shown that resorcin[4]arenes functionalised at their lower rim by azobenzenes **29-31** form well packed monolayers on water and also in LB films. These well packed films still allow the attached azobenzene groups to undergo reversible photoisomerisation from E to Z.[16,20] Isomerisation from E to Z in a water supported monolayer is accompanied by a corresponding increase in surface area and represents a drastic change in the shape of the azobenzene chromophores.

29 X = H,   R = (CH$_2$)$_3$CH$_3$
30 X = CH$_2$CO$_2$H, R = H
31 X = CH$_2$CO$_2$H, R = cyclohexyl

In 1994 reports appeared from the groups of Göpel and Reinhoudt,[21] Reinhoudt[22] and Stirling[23] on the formation of self-assembled monolayers of resorcin[4]arene-derived compounds through the formation of stable Au-S bonds. Göpel and Reinhoudt et al. report the self-assembly of a methyl bridged cavitand with four tetradialkyl sulfide legs on to a gold surface.[21] Importantly, the monolayer was shown by XPS experiments to be orientated with the sulfur atoms next to the gold surface and the oxygen atoms away from the surface. Reinhoudt et al. reported that the formation of the these monolayers at room temperature produced kinetically disordered monolayers, but heating to 60 °C promoted the formation of a well packed highly ordered structure of the calculated molecular dimensions (Figure 6).[22]

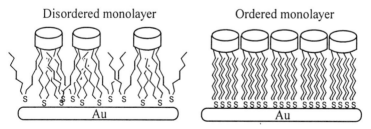

Figure 6. Schematic representation of a disordered monolayer of cavitand **34** and an ordered monolayer of cavitand **34** deposited at 60°C.

These highly ordered monolayers have been shown to form a surface hexagonal packing arrangement with a distance of 11.7 Å between centers.[24] Reinhoudt et al. also examined the effect of head group size on monolayer formation.[25,26] A variety of derivatives **32-40** were made and it was observed that well packed monolayers were formed only when the area of the head group of the resorcin[4]arene cavitand was equal to or smaller than that of the alkyl chains e.g. **32, 34, 38,** and **42**. Derivatives with head groups larger than the alkyl chains e.g. **33, 35, 36,** and **39** produced monolayers that had less densely packed alkyl chains. Strong hydrogen bonding between the free hydroxyl groups of resorcin[4]arene **40** lead to the formation of poorly organized bilayers.

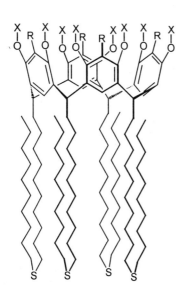

Resorcin[4]arenes

32	R = H	X = CH_3
33	R = OCH_3	X = CH_3
34	R = H	X = $C(O)CH_3$
35	R = CH_3	X = $C(O)CH_3$
36	R = $OC(O)CH_3$	X = $C(O)CH_3$
37	R = H	X = $CH_2C(O)OCH_3$
38	R = H	X = CH_2CH_2F
39	R = H	X = $CH_2C(O)NH(CH_2)_3H$
40	R = H	X = H

Cavitands

41	R = H	X = -CH2-
42	R = CH3	X = -CH2-

The length of terminal alkyl sulfide chains was also shown to effect monolayer formation, the best monolayers were formed when the attached dialkyl sulfide groups had a terminating alkyl chain of equal length to the primary alkyl chain.[25,26] The use of dialkyl sulfide chains to afford an Au-S linkage was extended to calix[4]arene and *p-tert*-Bu-calix[4]arenes derivatives which also gave well ordered monolayers.[27]

Similarly, Stirling et al. have demonstrated the formation of self-assembled gold-thiol monolayers from a resorcin[4]arene tetra thiol **26**.[23,28] Interestingly, when this layer was exposed to a solution of resorcin[4]arene **13** in hexane, the spontaneous formation of multilayers (up to 40) of resorcin[4]arene **13** on top of the original monolayer was observed.[28,29] Contact angle, grazing angle FTIR, X-ray and neutron reflectometry experiments strongly suggest that the multilayering structure takes the form of alternating layers of resorcin[4]arenes formed through bowl to bowl hydrogen bonding and deep interdigitation of the alkyl chains maximizing van der Waals contacts (Figure 7).[28,29] Interdigitation was also seen in the X-ray crystal structures of **13** and related compounds.[23,30] The multilayering phenomenon was shown to be sensitive to the solvent used, e.g. no multilayering was observed when ethanol was used as solvent. Indeed if a preformed multilayer was immersed in ethanol, the mulltilayer was dispersed and the monolayer restored. The stability of the multilayer could be considerably enhanced by the use of resorcin[4]arene **14** with alkene-functionalised legs followed by subsequent UV irradiated polymerization. Crosslinking the alkenyl legs helped to stabilize the multilayer towards solvation by ethanol.

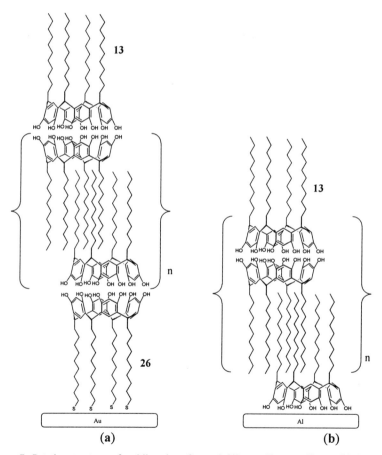

Figure 7. Putative structures of multilayering of resorcin[4]arene **13** on a self-assembled monolayer of resorcin[4]arene **26** on gold (**a**) and on an aluminium surface (**b**).

Stirling et al. have also shown that spontaneous self-assembled multilayering (7-40 layers) of resorcin[4]arene **13** occurs on substrates such as gold, stainless steel quartz, NaCl, monolayers of resorcin[4]arene **26**, 10-hydroxy-decanethiol and aluminum (Figure 7b).[31] Multilayering does not occur, however, on silicon and GaAs substrates.[31]

Ichimura et al. have recently observed the formation of monolayers of resorcin[4]arene **13** and its O-carboxymethoxylated (**43** and **44**) and O-hydroxyethoxylated (**45** and **46**) derivatives on suitable hydrophilic surfaces such as silica (i.e. quatrz) and poly-vinyl-alcohol, (Figure 8).[32] The authors demonstrated that surface adsorption was a reversible process that enabled formation of densely packed monolayers. Surface adsorption from highly dilute solutions could be accelerated by increasing the solution temperature. Desorption from the surfaces was dependent on the nature of the solvent with polar solvents (e.g. methanol) promoting desorption.[32]

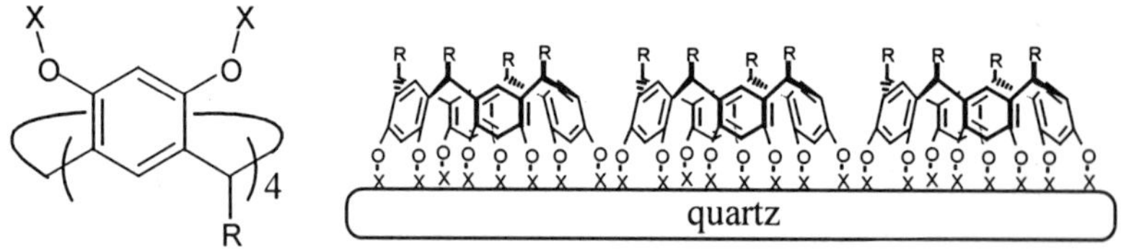

43 R = $(CH_2)_{10}CH_3$ X = CH_2CO_2H **45** R = $(CH_2)_4CH_3$ X = CH_2CO_2H
44 R = $(CH_2)_6CH_3$ X = CH_2CO_2H **46** R = $(CH_2)_{10}CH_3$ X = CH_2CH_2OH

Figure 8. Formation of densely packed monolayers of resorcin[4]arenes on a quartz surface.

The same workers have similarly adsorbed a resorcin[4]arene with four p-butylazobenzene groups **29** on to colloidal silica (Figure 9).[33] Aggregation properties of the coated colloidal silica in solution could be influenced through UV irradiation which resulted in E to Z isomerisation of the azobenzene units. Initially, the coated silica particles form very small aggregates if at all. However, following isomerisation to the Z form of the azobenzene unit much larger porous silica aggregates are formed. This was thought to be due to the increased polar nature of the Z-azobenzene group. The aggregation was also shown to be reversible through Z to E isomerisation.[33] These materials may form the basis for new materials with photo reversible dispersion-sedimentation cycles.

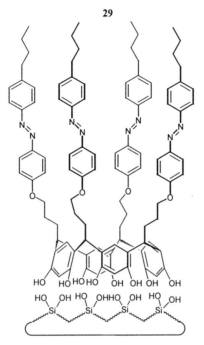

Figure 9. Illustrative representation of the azobenzene functionalised resorcin[4]arene **29** self-assembled on a colloidal silica surface.

Aoyama et al. have observed self-assembly of monolayers from octa-galactose functionalised resorcin[4]arenes onto quartz plates from an aqueous environment. These monolayers were stable in the aqueous environment and are discussed further later (*2.1 Adsorption*).[34,35]

8.2.1 Adsorption

The use of thin films for the adsorption of guest molecules has become an important aspect of applied calix[n]arene chemistry. Calix[n]arenes have been most frequently used as materials able to adsorb some ions or organic molecules. These have received considerable attention elsewhere.[2,21,36,37] Use of calix[n]arenes in thin films for the adsorption of guest molecules has, by comparison, received much less attention. Reports of adsorption by calix[n]arene thin films are often accompanied by ideas for practical applications as molecular or ion sensors.

We consider first the adsorptive properties of organized thin films prepared using self-assembly or LB techniques. These techniques produce thin films that allow the examination of calix[n]arenes in well ordered mono- and multilayers.

Mono- and multilayers of calix[n]arenes are, in fact two and three dimensional supramolecular arrays. These arrays may offer adsorption properties differing from those seen in solution. In fact, often they offer important insights in to molecular interactions and recognition at biological or sensor surfaces.

8.2.1.1 Adsorption of Metal and Molecular Ions

Although a great deal of work has examined the binding of metal ions by calix[n]arenes in solution,[2] adsorption by thin films at interfaces has received scant attention. Shinkai et al. were the first to report on the interactions of monolayers of ester-derived *p-tert*-Bu-calix[n]arenes with metal ions in the water subphase (Figure 10).[11] Pressure-area (π-A) isotherm curves for the calix[n]arenes on pure water and on metal salt solutions were compared. This revealed selectivity in the order $Li^+ < Na^+ > K^+ > Rb^+$ for **47**, $K^+ > Rb^+ > Na^+ > Li^+$ for **48**, and $Rb^+ > K^+ > Na^+ > Li^+$ for **49** in good agreement with selectivity experiments in solution.[11]

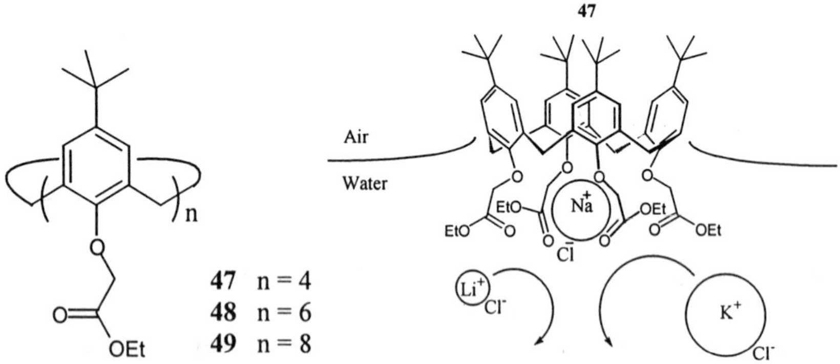

Figure 10. Representation of the selective formation of the calix[4]arene **47**/Na^+ complex at the air-water interface.

Stirling et al. have shown that the calix[8]arenes **50-52** form stable monolayers that may be trasfered to suitable substrates to form LB multilayers.[38] Such multilayers (16-fold) of the calix[8]arenes on silicon wafers were shown to adsorb a range of anions and cations from solution. For anions, the order of selectivity was I > F > Br > Cl and for cations was Cs > Rb > Na >> K as determined by XPS measurements.[38]

50 X = H
51 X = CH$_2$CO$_2$CH$_2$CH$_3$
52 X = CH$_2$COCH$_3$

Baglioni and coworkers have shown that a monolayer of *p-tert*-Bu-calix[6]arene **61** at the water-air interface shows high selectivity for Cs$^+$ over Na$^+$ and K$^+$ from the water subphase.[39] Shinkai et al. have also shown monolayers of calix[n]arenes at the air-water interface, complex with lanthanide ions in the subphase.[40]

Two interesting examples of ion binding and transport by calixarenes have recently been reported. Reinhoudt et al. has demonstrated that colloidal silica coated with an organic layer containing resorcarene **53** (Figure 11) shows affinity towards Cs$^+$.[41] Modified, selective collidal silica particles have particular application in the fields of separation and transport.

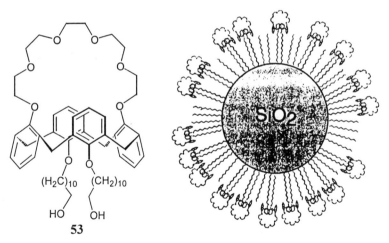

Figure 11. A sketch of a colloidal silica particle containing **53** as part of an cesium cation selective oganic coating.

Kobuke et al. have shown transport of sodium and potassium ions through a lipid bilayer containing a resorcin[4]arene containing C$_{17}$ *n*-alkyl chains.[42] A three fold increase in discrimination of potassium over sodium ions was measured, indicating the resorcarene is a synthetic peptide ion channel mimic.[42]

Amphiphilic [(4-alkylphenyl)azo]-substituted calix[n]arenes **54-57** form stable compressed monolayers supported on a buffered subphase.[43] The calix[n]arene monolayers were exposed to ammonium salts in the water subphase before and after compression. Phenyl and naphthyl ammonium salts were shown to bind differently to the monolayers when exposed to the monolayer before compression while a tetramethyl ammonium salt did not bind at all prior to compression. None of the ammonium salts were found to bind to precompressed monolayers.[43]

54 n = 4, R = $CH_2CH_2CH_2CH_3$
55 n = 4, R = $CH_2(CH_2)_6CH_3$
56 n = 6, R = $CH_2CH_2CH_2CH_3$
57 n = 6, R = $CH_2(CH_2)_6CH_3$

8.2.1.2 Adsorption of Neutral Organic Molecules

In 1991, Aoyama and co-workers examined the binding affinity of a series of sugars towards monolayers of resorcin[4]arene **13** supported on water a subphase or glass electrode.[13] The resorcin[4]arene monolayers provided a defined interface between the water subphase and the organic portion of the resorcin[4]arene monolayer layer. The affinity of sugars for the resorcin[4]arene thin film modified electrode increased in the order glucose < fucose ~ galactose ~ arabinose < xylose < ribose. Interestingly, the order was different from that observed for extraction experiments from the aqueous phase into CCl_4 where the order of affinity was xylose ~ galactose ~ glucose < arabinose < ribose < fucose.[13] The authors suggested that the monolayer of resorcin[4]arene **13** allows ribose to maximize both hydrophilic and hydrophobic interactions at the interface.

Recently, Aoyama has examined immobilized (as a monolayer on a quartz surface) amphiphilic resorcin[4]arene derivatives with either galactose **58** or glucose terminated side chains as mimics of naturally occurring cell-surface oligosaccharides.[44] These sugar cluster surfaces were effectively able to bind lectins, polysaccharides and poly(vinyl alcohol) in water. Further, Aoyama has shown that these macrocyclic sugar clusters are effectively able to bind 8-anilinonaphthalene in solution and then transport the guest to a quartz surface where it was trapped between the surface and the surface binding macrocyclic sugar cluster (Figure 12).[35] Aoyama and coworkers have also immobilised these macrocyclic sugar cluster resorci[4]arenes (through their hydrophobic alkyl chains) on hydrophobic sensor chips to observe the binding of lectins, polysaccharides and poly(vinyl alcohol) from the bulk aqueous phase.[44]

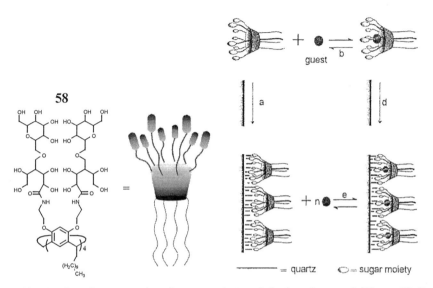

Figure 12. A schematic representaion of an octa-galactose derivative of a resorcin[4]arene **58**; its formation of self-assembled monolayers on a quartz surface (**a**), binding of a guest (**b**), transport of the bound guest to a quartz surface (**d**) and the reversible nature of binding of the guest at the surface (**e**).

Stirling et al. examined the selective adsorption of dilute aqueous solutions of a series of poly oxygenated compounds by self-assembled monolayers of a tetra thiol-resorcin[4]arene **26** on gold.[23] The monolayer was anchored by tetrapodal legs through a gold-thiol linkage. This structure provided a surface containing the bowl cavities as the molecular surface facing to the aqueous environment. The authors were able to show reversible binding for glucoronolactone, glutaric acid, butyrolactone and poly(vinyl pyrollidone). However, aqueous washout did not occur for glucoronolactone, the sodium salt of gluconic acid and vitamin C. The authors suggested strong multiple hydrogen bonding to the cavity to be responsible for this effect and formation of a covalent linkage in the case of vitamin C. This study was followed by a report examining a series of hydroxy lactones. The hydroxy lactones were shown to bind to the monolayers with ring fission of the lactone ring and acylation of the phenolic position.[28]

Self-assembled multilayers (7-40 layers, with a hydrophobic surface) of resorcin[4]arene **13** were shown to bind derivatives of glutaric acid[28,31] reversibly but did not bind vitamin C or glucoronolactone.[31]

Binding of organic amines from aqueous solution by thin films of a calix[4]arene-crown ether and calix[6] hexaester were observed using QCM sensors by Li et al.[45] A calix[6]arene hexaester provided a cavity capable of binding primary amines through tripodal hydrogen bonding to the ester carbonyl groups lining its cavity. The selectivity of the calix[6]arene hexaester towards alkylamines was in the order propylamine > butylamine > α-phenylethylamine > heptylamine ~ hexylamine > sec-butylamine >> triethylamine> dicyclohexylamine ~ diisobutylamine. Selectivity was based on the order of steric bulk of the amine.

Recently, Langmuir monolayers of chiral resorcin[4]arenes containing (S)(-)-phenylethylamine and (1R, 2S)(-)-norephedrine respectively were formed on a water subphase. The monolayers were able to discriminate between the chiral amino acids, alanine, valine, leucine and tryptophan through their π-A isotherms.[46]

Baglioni and coworkers have shown that Langmuir monolayers formed by spreading equimolar solutions of *p-tert*-Bu-calix[8]arene **62** and C_{60} on a water subphase exhibit the same π-A isotherms as a Langmuir monolayer of the **62**/C_{60} complex.[47] This result suggested the formation of stable **62**/C_{60} complexes at the air-water interface.

The adsorption of small neutral organic molecules by thin films of calix[n]arenes from the gas phase has received considerable attention from a number of research groups. Generally these groups have utilized the calix[n]arene thin films as potentially selective sensor coatings.

59 n = 4, R = H
60 n = 4, R = C(CH$_3$)$_3$
61 n = 6, R = C(CH$_3$)$_3$
62 n = 8, R = C(CH$_3$)$_3$
63 n = 4, R = CH(CH$_3$)$_2$

Göpel et al. examined the adsorption properties thin films of calix[n]arenes deposited on quartz crystal microbalances towards a variety of simple volatile organic molecules.[48,49] Films of thickness 5-150 nm were prepared either by dipping of substrates into corresponding solutions and subsequent evaporation of solvent or more generally, through the thermal evaporation of calix[n]arenes onto substrates via ultra-high vacuum ("Knudsen") conditions. Particular attention was paid to the response of a series of calix[n]arenes to vapours of perchloroethylene C_2Cl_4. The thin film mass change for different calix[n]arene thin films upon exposure to C_2Cl_4 increased in the order; **62** > **63** > **60**. It was observed that an increase in sensitivity occurred with increasing film thickness and that films also showed signs of aging effects (incomplete recovery over time). Differing vapour selectivity patterns were observed for **60** and **59** as determined by their binding equilibrium constants. Vapour binding selectivity increased in order C_6H_6 < $CHCl_3$ < C_2Cl_2 < $CH_3C_6H_5$ for **60** and in the order $CH_3C_6H_5$ < $CHCl_3$ < C_2Cl_2 for **59**.

Selectivity results matched quite closely with those determined theoretically from binding energies.[49] For C_2Cl_2 it was shown that there was an increased concentration of C_2Cl_2 at surface sites compared to bulk sites in the films. This effect was further confirmed through a study of the surface and bulk interactions between calix[n]arene films and organic vapours.[50] It was thought that this effect was due to fast surface adsorption into molecular cavities and a subsequently slower diffusion process to the less accessible bulk molecular cavities.

Independently, Dickert et al. also examined the response of thin films (40 nm) of a variety of calix[n]arenes on QCM sensors towards C_2Cl_4.[51] Their results also indicated the same trend in sensitivity relative to the size of the calix[n]arene macrocyclic as Göpel observed. Furthermore, Dickert observed that the inclusion of bulky trimethylsilyl groups at the lower rim of **60** dramatically reduced response to nearly negligible levels. Steric effects of the trimethylsilyl groups force the axes of the aromatic units into a parallel configuration causing the *tert*-butyl groups to close off access to the cavity. This result also showed that in this case at least extracavity (i.e. bulk) absorption by C_2Cl_4 was insignificant.[51]

Dickert et al. extended these results by comparing the response of resorcin[4]arenes and cavitands (similar to cavitand **58**) towards halogenated hydrocarbons (e.g. C_2Cl_4, CCl_4, $CHCl_3$ and CH_2CL_2) and aromatic compounds (e.g. C_6H_6, $C_6H_5CH_3$, *p*-xylene and *m*-xylene).[52,53] The effect of changing the size of attached alkyl chain on adsorption was also examined. The results indicated that the size of the attached alkyl chain did not influence the complexation properties but deep cavity cavitands selectively discriminated between xylene isomers based on their shape.[52] In a further study that correlated both theoretical and experimental results, Dickert and Schuster showed that thin films of **61** gave a large response to toluene vapours.[54] This was thought to be due to CH-π interactions between the host and guest respective alkyl and aromatic sites. The results also showed that the selective recognition of analytes by the thin films was due mainly to host/guest inclusion interactions.[54] In a later report Dickert et al. reports the detection of solvent vapours at concentrations as low as 2.5 ppm with a three sided quinoxaline-resorcin[4]arene cavitand fitted with long chain aliphatic spacers (similar to cavitand **65**).[53]

To more closely examine binding at surface sites, Göpel and Reinhoudt et al. reported binding experiments using a molecular monolayer formed from the resorcin[4]arene based cavitand **41** containing four dialkylsulfide chains (Figure 13).[21] This cavitand forms well ordered monolayers at gold surfaces by self-assembly to form stable Au-S linkages. Surface orientation is with chains perpendicular to the surface with the enforced head group cavity at the exposed surface.[21,55] In this position the cavity is in the perfect location for binding experiments.

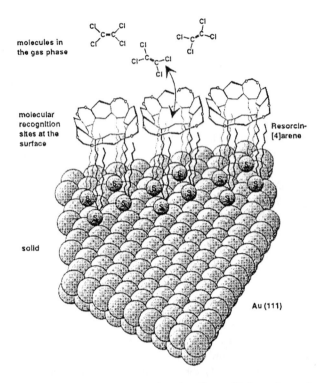

Figure 13. Schematic representation of the interaction of a self-assembled monolyaer of cavitand **41** on gold with tetrachloroethane in the gas phase.

Indeed when exposed to organic vapours the monolayer showed selective binding as determined by relative binding energies, increasing in the order $CCl_4 \sim CHCl_3 < C_2HCl_3 \sim CH_3C_6H_5 \ll C_2Cl_4$. The largest response and highest selectivity was for C_2Cl_4. The activation energy of desorption of C_2Cl_4 from the cavitand monolayer was similar to that observed for calix[4]arene multilayers described earlier. In comparison, self-assembled monolayers of didecylsulfide and octadecane thiol gave very much smaller responses to vapours of C_2Cl_4 presumably due to the lack of surface binding cavities.[55,56] A comparison of the response of the sulfide cavitand and octadecanethiol monolayers to organic vapours was measured by SPR spectroscopy.[24,57] This indicated that each of the above vapours generally produced a response which was at least a factor of two greater for the cavitand monolayer. The only exception was C_2Cl_4 which produced a response twelve times greater for the cavitand monolayer. This result further supports guest/host molecular recognition of C_2Cl_4 by the cavitand layer.[24]

Crookes et al. have demonstrated that *p-tert*-Bu-calix[4]arene **60** and *p-tert*-Bu-calix[6]arene **61** attached to a reactive monolayer surface increased the adsorption of organic vapours relative to control surfaces (Figure 14).[58] Selectivity between the calix[n]arenes towards the vapours was also observed. The authors concluded that these results strongly suggested an interaction between the calix[n]arene cavities and the organic vapours rather than nonspecific absorption.[58]

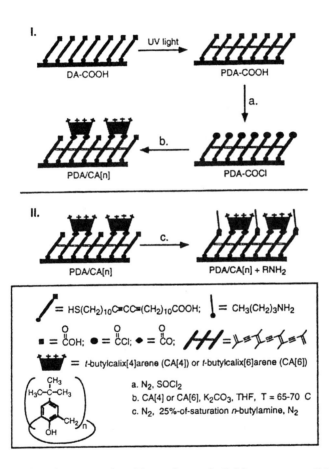

Figure 14. Schematic representation of the attachment of calix[n]arenes to a crosslinked reactive monolayer and the blocking of any remaining reactive sites with butylamine. The modified surface thus only allows interaction with volatile organic vapours through the calix[n]arene cavities in the surface.

Other extended cavitands have been shown to bind guests in the gas phase[59] and have been used as thin film sensors using QCM transducers.[60] In 1993, a cavitand based on a resorcin[4]arene bridged about the upper rim by four quinoxaline moieties was used to coat a QCM device with a thin film.[60] This thin film was shown to respond to organic vapours in the order $CCl_4 < C_6H_5F \sim$

$C_6H_5CH_3 \sim C_6H_6 \ll C_6H_5NO_2$. The film had a high selectivity for $C_6H_5NO_2$ as evidenced by the selectivity ratio for $C_6H_5NO_2 / C_6H_6$ which was 65. The authors suggested the greater sensitivity was due stronger π–π and dipole-dipole interactions between the cavitand and guest.[60] The cavitand films also showed no cross sensitivity to 90% relative humidity and gases such as CO, SO_2, NO, NO_2, CH_4, and C_3H_8.[60]

In similar experiments, Hartmann et al. also examined the response of extended quinoxaline-derived cavitands **64-66** with organic vapours and solutions containing organic pollutants.[61-63] Thin films of the cavitands were spray coated onto QCM surfaces which were then exposed to a series of aromatic (e.g. C_6H_6) and chlorinated vapours (e.g. CCl_4). A comparison of binding properties of the vase and kite conformations of cavitand **64** (Figure 15) revealed the kite structure to have reduced sensitivity towards the organic vapours. Also the relative selectivity of the kite was reduced and differed from that of the vase, preferring the more polar analytes (e.g. methanol, ethanol, diethylamine and acetonitrile). These effects were thought to be due to the lack of a preorganized cavity and the exposure of the nitrogen and oxygen atoms of the kite structure leading to increased polar interactions.[61]

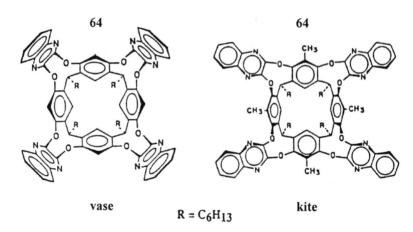

Figure 15. Representation of the vase and kite forms of cavitand **64** respectively.

A later paper examined the influence of the vase cavity dimensions on adsorption of organic vapours.[62] It was shown that the sensitivity and selectivity of the binding response varied with the size and shape of both upper and lower rim substituents. Cavitand **64** gave a good responses to $CHCl_3$, CCl_4 and to a lesser extent C_6H_6 with other non aromatic and polar molecules giving smaller responses. However, the larger deeper cavities of **64** and **65** had reduced selectivity compared to shallower derivative **67** that allowed access to the electron rich cavity by molecules able to engage in CH-π interactions (e.g. nitromethane, ethyl acetate and acetonitrile).

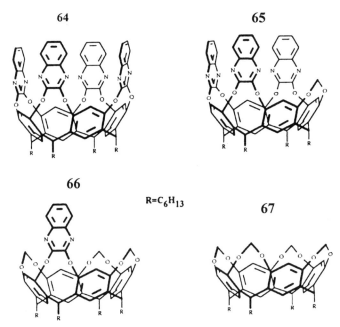

The effect of the changing R groups at the lower rim produced increased selectivity when longer alkyl chains were attached. The authors suggest this is due to the formation of dispersed and porous layers.[62] Hartmann and coworkers have shown that thin films of **62** are able to detect $CHCl_3$ in solution at 200 ppm concentration.[63]

Multilayers of Langmiur-Blodgett (LB) films of resorcin[4]arenes on QCM devices have been shown to absorb saturated vapours of C_6H_6, $C_6H_5CH_3$ and aniline. The films showed shown no selectivity at these concentration levels with absorption occurring in the film bulk with up to 20 vapour molecules per molecular cavity.[15] Similar results were obtained for LB multilayers of phosphorylated resorcin[4]arenes studied by SPR.[17] The SPR response of the film to high concentrations of $C_6H_5CH_3$ and petrol vapours indicated that films suffered from swelling. This was explained by the film capturing and subsequently condensing the guest molecules in the film matrix.[17] Changes in film thickness and optical properties of thermally evaporated films of *p-tert*-Bu-calix[n]arenes have also been observed following absorption of aromatic guests into the films.[64]

8.2.2 Nonlinear Optical Compounds

Interest in the nonlinear optical (NLO) properties and behaviour of calix[n]arene derivatives is an emerging field of interest. NLO compounds are of interest for their use in devices that have applications like optical switching, communications and the frequency doubling of low cost laser sources which can be used to achieve higher data storage capacities. NLO organic compounds are generally polarised (donor-acceptor molecules linked by π conjugation) conjugated molecules that are noncentrosymmetric. Calix[n]arenes as donor-acceptor aromatic molecules are often found in the cone conformer, a noncentrosymmetric molecular arrangement.

Reinhoudt et al. first reported a series of calix[4]arenes containing *p*-nitro substituents and groups bearing nitro substituents **68-73**.[65,66] These molecules displayed NLO properties in solution and so their macroscopic properties were examined through the use of thin films. The films were spin cast onto appropriate substrates and then subjected to poling by a strong electric field. These materials formed stable films with good optical properties that were suitable for waveguide type applications (e.g. optical sensors). Reinhoudt and coworkers have also recently shown that thin poled films of **68** form two distinct molecular crystalline domains of either a rectangular or hexagonal type arrangement.[67,68]

68 $R^{1-4} = NO_2$ $X^{1-4} = n\text{-Pr}$
69 $R^1 = NO_2, R^{2-4} = H$ $X^{1-4} = n\text{-Pr}$
70 $R^{1,3} = NO_2, R^{2,4} = H$ $X^{1-4} = n\text{-Pr}$
71 $R^{1,2} = NO_2, R^{3,4} = H$ $X^{1-4} = n\text{-Pr}$
72 $R^{1-4} = (E)CH=CHC_6H_4NO_2$ $X^{1-4} = n\text{-Pr}$
73 $R^{1-4} = (E)N=NC_6H_4NO_2$ $X^{1,3} = H, X^{2,4} = n\text{-Pr}$

Another interesting recent development by Reinhoudt et al is the synthesis of NLO active calix[4]arene polyimides **74**.[69] This calix[4]arene polyimide forms thin films that are smooth, physically stable and are highly transparent. These features are all important for future practical applications.

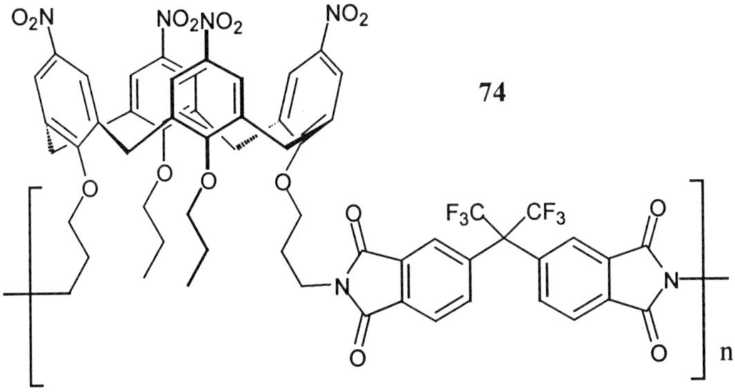

74

Morley and Naji have reported a theoretical (AM1) study on a series of calix[n]arenes which examined structural effects on dipole moments and hyperpolarisabilities of calix[n]arenes.[70] They found that all the molecules examined were expected to have large dipole moments but that their hyperpolarisabilities were dependant on the number of molecular excited states included and on the orientation of the donor substituents. Increases in the calix[n]arene ring size caused only small effects on their electronic properties.

Li et al. have recently reported on a self-assembled monolayer of calix[n]arene-based NLO molecular "pyramids" (Figure 16).[71,72] The surface bound calix[4]arene provides a densely packed, highly ordered monolayer where noncentrosymmetric orientation within the film is maintained by covalent binding to the silica surface. The monolayer structure exhibited a charge transfer band at 380 nm that was red-shifted from the electronic absorption spectrum in solution and displays extremely large second-order nonlinearities with a robust molecular dipole alignment.[71,72]

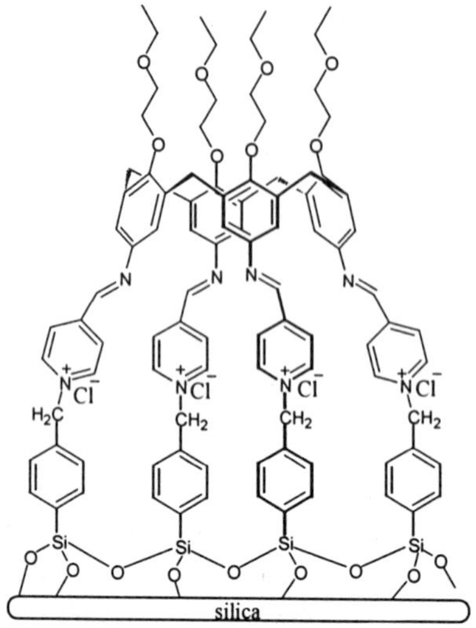

Figure 16. A schematic representation of a self-assembled monolayer of a calix[4]arene pyramid on a silica surface for NLO applications.

8.2.3 Pyroelectric Effects

The pyroelectric effect is a temperature-dependent electric polarisation of compounds or materials containing a noncentrosymetric structure. Recently Richardson et al. reported for the first time an LB mixed multilayer material containing calix[n]arenes.[19] An LB multilayer using carboxylic acid and amine functionalised calix[n]arenes, **75** and **76** respectively was constructed to examine the pyroelectric properties of the films.

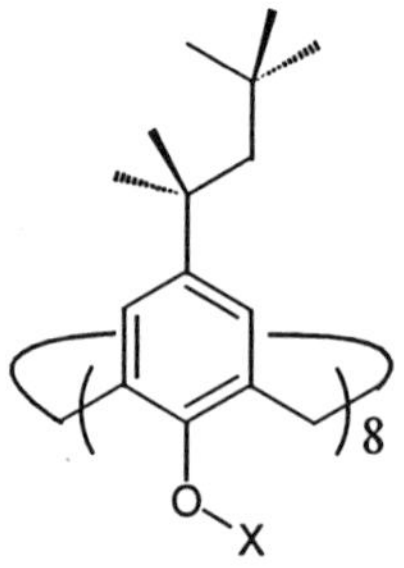

75 X = CH_2CO_2H
76 X = $CH_2CH_2CH_2NH_2$

A multilayer (11 layers) was built up in an alternating **75-76-75-76** pattern on an aluminum surface and gave a high pyroelectric coefficient of ~7 $\mu Cm^{-2}K^{-1}$ when compared to other fatty acid / fatty amine LB films.[19]

8.2.4 Luminescence and Fluorescence of Monolayers

Metal complexes of calix[n]arenes and suitably functionalised calixarenes have been known for some time to display luminescence properties in solution.[2] However, there are few reports of these materials being utilised in thin films. Shinkai et al. have demonstrated the luminescence and fluorescence of some metal complexed calix[n]arenes at the air water interface.[40] The authors demonstrated the fluorescence of monolayers of calix[4]arene **77** and **78** when complexed with lanthanide ions such as Sm^{3+}, Eu^{3+}, Tb^{3+} and Dy^{3+} from the water subphase. It was also shown that the fluorescence emission spectra could be tuned by changing the surface pressure applied to the monolayers.

77 R = *tert*-Octyl, R' = OH
78 R = *tert*-Butyl, R' = NHOH

Calix[n]arene carboxylic acids have also been used to coat porous silica to modify its electrochemiluminescence properties. The calix[n]arene carboxylic acid coating on the silica leads to lower electrochemiluminescence but longer performance time periods.[73]

8.2.5 Other Applications of Thin Films

Thin films of calix[n]arenes have been examined to determine their relevent properties for use as or in semiconductor type materials for electronic and sensor devices. Guillaud et al. have reported on electrical measurements of thin layers (1000 Å to 2 μm thick) of *p*-isopropyl-calix[6]arene on Al, Cr and Au electrodes.[74,75] The *p*-isopropyl-calix[6]arene was thermally evaporated (at 10^{-7} torr) on to an electrode and then thermally annealed. The films were found to be highly insulating materials but through the use of iodine doping, the films became conducting. The observed electrical activity was also shown to depend on the electrode metal and on the humidity (conductivity increases with humidity).[74,75] In later work Guillaud and coworkers examined the electrical properties of a metal-calix[n]arene-semiconductor structure.[76]

Another application of calix[n]arenes in semiconductor materials is their use in the formation of cadmium sulfide (CdS) nano-particles. Richardson et al. have demonstrated the formation of CdS nanoparticles within LB films of cadmium salts of **79** and **80** by reaction with hydrogen sulfide gas.[77]

[structure: calix[n]arene with CH₂CO₂H group]

79 n = 4
80 n = 8

The formation and size of the cadmium sulfide nano-particles was confirmed using UV-vis absorption spectra and XPS measurements. The size of the nano particles was 1.5 ±0.3 nm and did not depend on the type of calix[n]arene or the number of LB layers but increased with the pH of the subphase used for deposition of the LB layers.[77]

Reinhoudt and coworkers have examined as self-assembled monolayers of a carceplex-DMF complex on a gold surface as a molecular switching device (Figure 17). Within the carceplex framework the bound guest DMF molecule can switch (via rotation) between two differing orientations.[24]

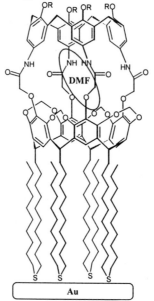

Figure 17. Schematic representation of a self-assembled monolayer of a non-symmetrical carceplex containing a guest DMF molecule on gold. The guest DMF molecule can adopt two conformations within the carceplex and demonstrates the possibility of a molecular switching device.

Such switching devices, when used in digital data storage materials, could greatly increase storage capacity by utilising molecular sized switches.

8.3 Conclusions

Calix[n]arenes and their related macrocycles have undergone a considerable amount of fundamental research on their ability to form ordered supramolecular thin films. They have been used successfully in a wide range of thin film deposition techniques e.g. self-assembly, Langmuir-Blodgett, and evaporative methods. The range of techniques available is due to the easy modification of calix[n]arenes to suit each deposition technique. Once formed, the calix[n]arene thin films have been shown to be sensitive to a range of metal ions, molecular ions and neutral organic molecules through interactions with their cavities. These processes can be exploited for sensor and biological applications. Furthermore, thin films of some calix[n]arenes have been shown to have NLO, pyroelectric, and semiconductor type properties. Thin films of calix[n]arenes and their related compounds have only just begun to yield exciting advances in the area of ordered supramolecular materials, thus, future research in this area looks very promising.

8.4 References

1. Gutsche, D. C. *Calixarenes*; (The Royal Society of Chemistry: Cambridge, 1992).
2. Gutsche, C. D. *Calixarenes Revisited*; (The Royal Society of Chemistry: Cambridge, 1998).
3. Lhotak, P.; Shinkai, S. *J. Synth. Org. Chem. Jpn.*, **53** (1995), 963-974.
4. Higler, I.; Timmerman, P.; Verboom, W.; Reinhoudt, D. N. *Euro. J. Org. Chem.*, (1998), 2689-2702.
5. Markowitz, M. A.; Bielski, R.; Regen, S. L. *J. Am. Chem. Soc.*, **110** (1988), 7545-7546.
6. Markowitz, M. A.; Janout, V.; Castner, D. G.; Regen, S. L. *J. Am. Chem. Soc.*, **111** (1989), 8192-8200.
7. Conner, M.; Janout, V.; Regen, S. L. *J. Am. Chem. Soc.*, **115** (1993), 1178-1180.
8. Dedek, P.; Webber, A. S.; Janout, V.; Hendel, R. A.; Regen, S. L. *Langmuir*, **10** (1994), 3943-3945.
9. Lee, W.; Hendel, R. A.; Dedek, P.; Janout, V.; Regen, S. L. *J. Am. Chem. Soc.*, **117** (1995), 6793-6794.
10. Hendel, R. A.; Janout, V.; Lee, W.; Regen, S. L. *Langmuir*, **12** (1996), 5745-5746.
11. Ishikawa, Y.; Kunitake, T.; Matsuda, T.; Otsuka, T.; Shinkai, S. *J. Chem. Soc., Chem. Commun.*, (1989), 736-738.

12. Merhi, G.; Munoz, M.; Coleman, A. W.; Barrat, G. *Supramol. Chem.*, **5** (1995), 173-177.
13. Kurihara, K.; Ohto, K.; Tanaka, Y.; Aoyama, Y.; Kunitake, T. *J. Am. Chem. Soc.*, **113** (1991), 444-450.
14. Moreira, W. C.; Dutton, P. J.; Aroca, R. *Langmuir*, **11** (1995), 3137-3144.
15. Nabok, A. V.; Lavrik, N. V.; Kazantseva, Z. I.; Nesterenko, B. A.; Markovskiy, L. N.; Kalchenko, V. I.; Shivaniuk, A. N. *Thin Solid Films*, **259** (1995), 244-247.
16. Ichimura, K.; Fukushima, N.; Fujimaki, M.; Kawahara, S.; Matsuzawa, Y.; Hayashi, Y.; Kudo, K. *Langmuir*, **13** (1997), 6780-6786.
17. Nabok, A. V.; Hassan, A. K.; Ray, A. K.; Omar, O.; Kalchenko, V. I. *Sens. Actuators B*, **45** (1997), 115-121.
18. Davis, F.; Lucke, A. J.; Smith, K. A.; Stirling, C. J. M. *Langmuir*, **14** (1998), 4180-4185.
19. Richardson, T.; Greenwood, M. B.; Davis, F.; Stirling, C. J. M. *Langmuir*, **11** (1995), 4623-4625.
20. Fujimaki, M.; Matsuzawa, Y.; Hayashi, Y.; Ichimura, K. *Chem. Lett.*, (1998), 165-166.
21. Schierbaum, K. D.; Weiss, T.; Vanvelzen, E. U. T.; Engbersen, J. F. J.; Reinhoudt, D. N.; Gopel, W. *Science*, **265** (1994), 1413-1415.
22. Thoden van Velzen, E. U.; Engbersen, J. F. J.; Reinhoudt, D. N. *J. Am. Chem. Soc.*, **116** (1994), 3597-3598.
23. Adams, H.; Davis, F.; Stirling, C. J. M. *J. Chem. Soc., Chem. Commun.*, (1994), 2527-2529.
24. Huisman, B. H.; van Veggel, F. C. J. M.; Reinhoudt, D. N. *Pure & Appl. Chem.*, **70** (1998), 1985-1992.
25. Thoden van Velzen, E. U.; Engbersen, J. F. J.; Reinhoudt, D. N. *Synthesis*, (1995), 989-997.
26. Vanvelzen, E. U. T.; Engbersen, J. F. J.; Delange, P. J.; Mahy, J. W. G.; Reinhoudt, D. N. *J. Am. Chem. Soc.*, **117** (1995), 6853-6862.
27. Huisman, B. H.; Vanvelzen, E. U. T.; Vanveggel, F.; Engbersen, J. F. J.; Reinhoudt, D. N. *Tetrahedron Lett.*, **36** (1995), 3273-3276.
28. Davis, F.; Stirling, C. J. M. *Langmuir*, **12** (1996), 5365-5374.
29. Davis, F.; Stirling, C. J. M. *J. Am. Chem. Soc.*, **117** (1995), 10385-10386.
30. Hibbs, D. E.; Hursthouse, M. B.; Malik, K. M. A.; Adams, H.; Stirling, C. J. M.; Davis, F. *Acta Crystallog. Sec. C., Struct. Commun.*, **54** (1998), 987-992.
31. Davis, F.; Gerber, M.; Cowlam, N.; Stirling, C. J. M. *Thin Solid Films*, **285** (1996), 678-682.
32. Kurita, E.; Fukushima, N.; Fujimaki, M.; Matsuzawa, Y.; Kudo, K.; Ichimura, K. *J. Mater. Chem.*, **8** (1998), 397-403.
33. Ueda, M.; Fukushima, N.; Kudo, K.; Ichimura, K. *J. Mater. Chem.*, **7** (1997), 641-645.

34. Fujimoto, T.; Shimizu, C.; Hayashida, O.; Aoyama, Y. *Gazz. Chim. Ital.*, **127** (1997), 749-752.
35. Fujimoto, T.; Shimizu, C.; Hayashida, O.; Aoyama, Y. *J. Am. Chem. Soc.*, **119** (1997), 6676-6677.
36. Perrin, R.; Lamartine, R.; Perrin, M. *Pure Appl. Chem.*, **65** (1993), 1549-1559.
37. Diamond, D. *J. Inclusion Phenom. Mol. Recognit. Chem.*, **19** (1994), 149-166.
38. Davis, F.; Otoole, L.; Short, R.; Stirling, C. J. M. *Langmuir*, **12** (1996), 1892-1894.
39. Dei, L.; Casnati, A.; Lonostro, P.; Baglioni, P. *Langmuir*, **11** (1995), 1268-1272.
40. Ludwig, R.; Matsumoto, H.; Takeshita, M.; Ueda, K.; Shinkai, S. *Supramol. Chem.*, **4** (1995), 319-327.
41. Nechifor, A. M.; Philipse, A. P.; deJong, F.; vanDuynhoven, J. P. M.; Egberink, R. J. M.; Reinhoudt, D. N. *Langmuir*, **12** (1996), 3844-3854.
42. Tanaka, Y.; Kobuke, Y.; Sokabe, M. *Angew. Chem. Int. Ed. Engl.*, **34** (1995), 693-694.
43. Tyson, J. C.; Moore, J. L.; Hughes, K. D.; Collard, D. M. *Langmuir*, **13** (1997), 2068-2073.
44. Hayashida, O.; Shimizu, C.; Fujimoto, T.; Aoyama, Y. *Chem. Lett.*, (1998), 13-14.
45. Zhou, X. C.; Ng, S. C.; Chan, H. S. O.; Li, S. F. Y. *Sens. Actuators B*, **42** (1997), 137-144.
46. Pietraszkiewicz, M.; Prus, P.; Fabianowski, W. *Pol. J. Chem.*, **72** (1998), 1068-1075.
47. Dei, L.; LoNostro, P.; Capuzzi, G.; Baglioni, P. *Langmuir*, **14** (1998), 4143-4147.
48. Schierbaum, K. D.; Gerlach, A.; Haug, M.; Gopel, W. *Sens. Actuators A*, **31** (1992), 130-137.
49. Dominik, A.; Roth, H. J.; Schierbaum, K. D.; Gopel, W. *Supramol. Sci.*, **1** (1994), 11-19.
50. Schiebaum, K. D.; Gerlach, A.; Gopel, W.; Muller, W. M.; Vogtle, F.; Dominik, A.; Roth, H. J. *Fresenius J. Anal. Chem.*, **349** (1994), 372-379.
51. Dickert, F. L.; Schuster, O. *Adv. Mater.*, **5** (1993), 826-829.
52. Dickert, F. L.; Baumler, U. P. A.; Zwissler, G. K. *Synth. Metals*, **61** (1993), 47-52.
53. Dickert, F. L.; Baumler, U. P. A.; Stathopulos, H. *Anal. Chem.*, **69** (1997), 1000-1005.
54. Dickert, F. L.; Schuster, O. *Mikrochim. Acta*, **119** (1995), 55-62.
55. Weiss, T.; Schierbaum, K. D.; Vanvelzen, U. T.; Reinhoudt, D. N.; Gopel, W. *Sens. Actuators B*, **26** (1995), 203-207.
56. Rickert, J.; Weiss, T.; Gopel, W. *Sens. Actuators B*, **31** (1996), 45-50.
57. Huisman, B. H.; Kooyman, R. P. H.; vanVeggel, F.; Reinhoudt, D. N. *Adv. Mater.*, **8** (1996), 561-566.

58. Dermody, D. L.; Crooks, R. M.; Kim, T. *J. Am. Chem. Soc.*, **118** (1996), 11912-11917.
59. Vincenti, M.; Dalcanale, E.; Soncini, P.; Guglielmetti, G. *J. Am. Chem. Soc.*, **112** (1990), 445-447.
60. Nelli, P.; Dalcanale, E.; Faglia, G.; Sberveglieri, G.; Soncini, P. *Sens. Actuators B*, **13** (1993), 302-304.
61. Dalcanale, E.; Hartmann, J. *Sens. Actuators B*, **24** (1995), 39-42.
62. Hartmann, J.; Hauptmann, P.; Levi, S.; Dalcanale, E. *Sens. Actuators B*, **35** (1996), 154-157.
63. Hartmann, J.; Auge, J.; Lucklum, R.; Rosler, S.; Hauptmann, P.; Adler, B.; Dalcanale, E. *Sens. Actuators B*, **34** (1996), 305-311.
64. Shirshov, Y. M.; Zynio, S. A.; Matsas, E. P.; Beketov, G. V.; Prokhorovich, A. V.; Venger, E. F.; Markovskiy, L. N.; Kalchenko, V. I.; Soloviov, A. V.; Merker, R. *Supramol. Sci.*, **4** (1997), 491-494.
65. Kelderman, E.; Derhaeg, L.; Heesink, G. J. T.; Verboom, W.; Engbersen, J. F. J.; Vanhulst, N. F.; Persoons, A.; Reinhoudt, D. N. *Angew. Chem. Int. Ed. Engl.*, **31** (1992), 1075-1077.
66. Kelderman, E.; Heesink, G. J. T.; Derhaeg, L.; Verbiest, T.; Klaase, P. T. A.; Verboom, W.; Engbersen, J. F. J.; Vanhulst, N. F.; Clays, K.; Persoons, A.; Reinhoudt, D. N. *Adv. Mater.*, **5** (1993), 925-930.
67. Schonherr, H.; Kenis, P. J. A.; Engbersen, J. F. J.; Harkema, S.; Hulst, R.; Reinhoudt, D. N.; Vancso, G. J. *Langmuir*, **14** (1998), 2801-2809.
68. Kenis, P. J. A.; Noordman, O. F. J.; Schonherr, H.; Kerver, E. G.; SnellinkRuel, B. H. M.; vanHummel, G. J.; Harkema, S.; vanderVorst, C.; Hare, J.; Picken, S. J.; Engbersen, J. F. J.; vanHulst, N. F.; Vancso, G. J.; Reinhoudt, D. N. *Chem. Euro. J.*, **4** (1998), 1225-1234.
69. Kenis, P. J. A.; Noordman, O. F. J.; vanHulst, N. F.; Engbersen, J. F. J.; Reinhoudt, D. N.; Hams, B. H. M.; vanderVorst, C. *Chem. Mater.*, **9** (1997), 596-601.
70. Morley, J. O.; Naji, M. *J. Phys. Chem. A*, **101** (1997), 2681-2685.
71. Yang, X. G.; McBranch, D.; Swanson, B.; Li, D. Q. *Angew. Chem. Int. Ed. Engl.*, **35** (1996), 538-540.
72. Li, D. Q.; Yang, X. G.; McBranch, D. *Synth. Metals*, **86** (1997), 1849-1850.
73. Zhang, L. B.; Coffer, J. L. *J. Phys. Chem. B*, **101** (1997), 6874-6878.
74. Chaabane, R. B.; Gamoudi, M.; Guillaud, G.; Jouve, C.; Gaillard, F.; Lamartine, R. *Synth. Metals*, **66** (1994), 49-64.
75. Chaabane, R. B.; Gamoudi, M.; Guillaud, G. *Synth. Metals*, **67** (1994), 231-233.
76. Chaabane, R. B.; Gamoudi, M.; Remaki, B.; Guillaud, G.; El Beqqali, O. *Thin Solid Films*, **296** (1997), 148-151.
77. Nabok, A. V.; Richardson, T.; Davis, F.; Stirling, C. J. M. *Langmuir*, **13** (1997), 3198-3201.

CHAPTER 9

CALIXARENES IN SELF-ASSEMBLY PHENOMENA

VOLKER BÖHMER and ALEXANDER SHIVANYUK

9.1. Introduction

Design and investigation of molecules capable to form well defined, functional supramolecular structures by self-assembly has attracted considerable interest during the last decade. This is not only due to the great importance of such processes for biological systems. The development of artificial molecules bearing in their chemical structure all the informations to built up assemblies of increasing complexity is a challenge in itself.[1] Not simply a molecular structure is copied from Nature, like in the classical total synthesis of compounds found in living organisms. It is the building principle itself, the way of constructing highly sophisticated architectures which is transferred to new, hitherto unknown systems.

Self-assembly means connection via "reversible bonds" such as hydrogen bonds, charge transfer interactions, metal co-ordination, van der Waals and solvophobic forces. Most of these interactions are rather weak, as compared to covalent links. Therefore, a high stability of any envisaged self-assembled species can be reached only via a positive co-operation of such weak attractions. This reinforcement may be achieved by preorganization of several binding groups on a suitable molecular skeleton. Obviously, calixarenes and resorcarenes[2] belong to the most promising molecular platforms, since they are readily available and amenable to all kinds of chemical modifications allowing almost unlimited variations in the combination and the mutual geometrical arrangement of different functional groups. In addition, their concave shape makes them especially attractive for the construction of closed, container-like structures capable of molecular encapsulation.[3]

The following chapter concentrates on self-assembled supramolecular structures held together by hydrogen bonds, the most important "connection" in natural systems. Assemblies formed via solvophobic effects or by coordinative bonds are excluded. The main emphasis is given to association in solution, while self-assembled architectures in the crystalline state are not exhaustively treated.

9.2. Calixarenes

9.2.1. Calixarenes in the Cone Conformation

Calix[4]arene derivatives, fixed in the cone conformation may be functionalized at the narrow or at the wide rim by hydrogen bond donors or acceptors as well as by self-complementary hydrogen bonding fragments.

Early studies with the mono- and di-(2-pyridone) derivatives **1a,b** showed, that **1a** dimerizes in the same way as pyridone itself (K = 100 ± 20 M^{-1}), while **1b** gives oligomeric hydrogen bonded aggregates.[4] Obviously, the flexibility especially of the pendant groups prevents a clean dimerization. The aggregates could be destroyed ("denatured") by a complementary urea derivative (imidazolidine).

The formation of well-defined hydrogen bonded dimers was observed in the case of the calix[4]arene dicarboxylic acid **2**.[5] In the crystalline state **2a** exists in a C_{2v}-symmetrical pinched cone conformation in which the two aryl rings bearing the carboxylic groups are quasi-parallel. Two molecules **2a** form a centro-symmetrical dimer via four C=O···H-O hydrogen bonds. The corresponding pinched cone conformation which allows no guest inclusion was also found in the apolar solvents such as $CDCl_3$ and CD_2Cl_2 as the only species.

2 n = 2

a $Y^1 = Y^2 = C_3H_7$ $R = NO_2$
b $Y^1 = Y^2 = C_8H_{17}$ $R = H$
c $Y^1 = Y^2 = C_2H_4\text{-}O\text{-}C_2H_5$
     $R = H$

3 n = 3

$Y^1 = CH_3$, $Y^2 = CH_3$, C_8H_{17}
$R = C(CH_3)_3$

1a R = H

1b R = NH

In the case of the conformationally more flexible calix[6]arene tricarboxylic acid **3** the existence of dimers in apolar solvents was proved by VPO and ^1H NMR-spectroscopy.[6] While in CD_3OD only one broad singlet is observed for the methylene bridges two distinct doublets are found in $CDCl_3$ due to the rigidification of the flattened cone conformation in the dimer. The cavity thus formed is obviously large enough to include cations such as N-methylpyridinium and N-methyl-4-picolinium (but not N-methyl-2-picolinium) since significant upfield shifts were found for all signals upon addition of their iodides. Stability constants of 347 ± 8 and 234 ± 6 M^{-1}, respectively, were evaluated for these kinetically non stable complexes.

Carboxylic groups (as hydrogen bond donors) at the wide rim of calix[4]arenes were also used in combination with pyridine residues (as acceptors) attached either to the wide or narrow rim.

The tetrapyridyl derivative **5** solubilizes exactly one equivalent of the tetraacid **4a** in $CDCl_3$ suggesting that a capsule-like hetero-dimer is formed via four strong COOH···N hydrogen bonds.[7] Although the ^1H NMR-spectrum of **5** does not change considerably upon complexation with **4a** and the signal for carboxylic protons remains broad, the molecular weight (VPO) was consistent with the dimer formation. However, no molecular encapsulation was reported in this case.

Similarly the tetraacid **4b** could be solubilized in $CDCl_3$ by the 4-pyridyl derivative **6a** and its 3-pyridyl analogue **6b** in a 1:1 ratio while no solubilization took place in the case of the 2-pyridyl derivative **6c**.[8] The formation of the suggested complexes **4b·6a** and **4b·6b** was additionally confirmed by molecular weight determination (VPO). Stability constants of $7.6·10^3$ and $1.3·10^3$ M^{-1} were derived from dilution experiments by ^1H NMR (ArH signal of **6**). Two signals observed for the aromatic protons of **4b** in aggregates with *all* pyridine derivatives (even with 4-picoline) indicate a pinched cone conformation.

No interaction was found between the 1,3-dicarboxylic acid **2c** and tetra- or 1,3-dipyridyl derivatives, probably due to the dimerization of the self-complementary diacids **2** described above. On the other hand **4b** can be solubilized by sufficient amounts of 1,3-dipyridyl ethers analogous to **6a,b** (which are not self-complementary).

Several examples are known in which the hydrogen bonding properties of calixarene derivatives are controlled by conformational changes induced by metal ion complexation.[9] Calixarene **7** bearing two diamidopyridine fragments at the narrow rim does not interact via intermolecular hydrogen bonds with the complementary flavine derivative **8** due to strong intramolecular C=O···H-N hydrogen bonds. The complexation of a sodium cation disrupts the intramolecular hydrogen bonds so that an interaction of the diamidopyridine moieties of **7** with **8** becomes possible (Fig. 9.1.). Polymeric hydrogen bonded structures of calix[4]arenes capable of Na$^+$ complexation were also described.[10]

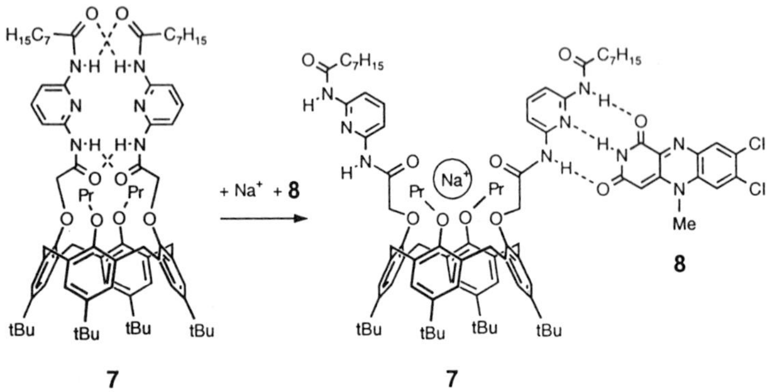

Figure 9.1. Na$^+$-controlled hydrogen bonding.

This principle was also used for the self-assembly of a bifunctional receptor[11] (Fig. 9.2.) from the Na$^+$-complex of **9** (the cation receptor) and a complementary thymine substituted Zn-porphyrin **10** (the anion receptor).

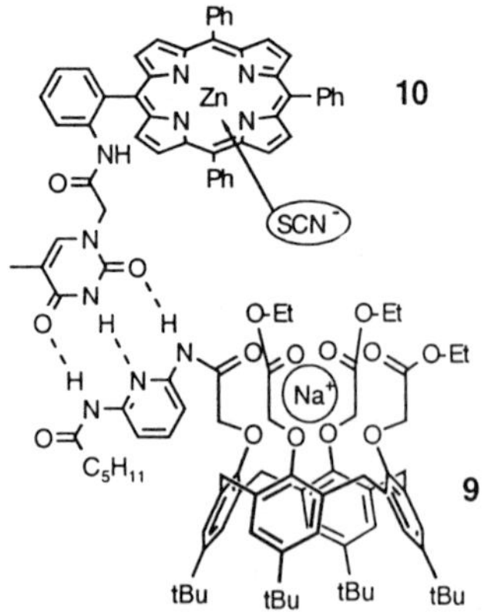

Figure 9.2. A self-assembling bifunctional receptor for NaSCN.

Most of these aggregates formed in solution are kinetically not stable and thus, only indirect conclusions about their structure can be drawn. Much stronger structural evidence has been accumulated (mainly from ^{1}H NMR) for the kinetically stable self-assembled aggregates described in the next sections.

9.2.2. Tetra-Urea Derivatives

Tetraaminocalix[4]arenes are readily available by (ipso-)nitration and reduction[12] or by Ullmann coupling of tetraiodo derivatives with phthalimide and subsequent hydrazinolysis.[13] Their reaction with isocyanates offers an easy access to a huge variety of tetra-urea derivatives **11** differing by the residues Y and R attached to the phenolic oxygens and to the urea groups.

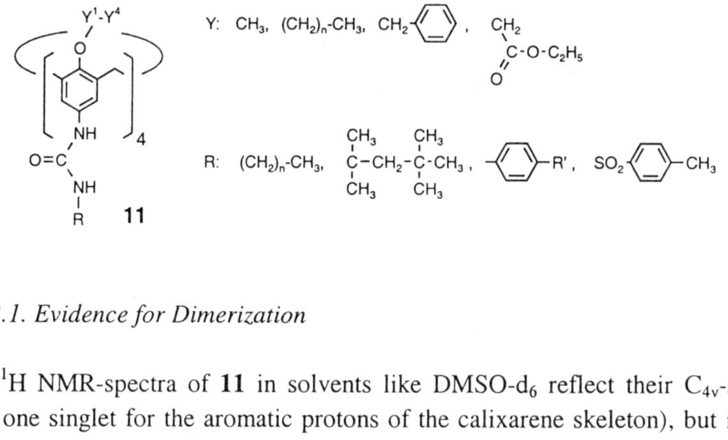

9.2.2.1. Evidence for Dimerization

The ^{1}H NMR-spectra of **11** in solvents like DMSO-d_6 reflect their C_{4v}-symmetry (e.g. one singlet for the aromatic protons of the calixarene skeleton), but indicate a less symmetric pattern in apolar solvents such as CDCl$_3$ or benzene-d_6 (e.g. a pair of m-coupled doublets for these aromatic protons). For an example see Fig. 9.3. In addition, one of the NH-singlets is strongly downfield shifted, typical for strong hydrogen bonding.

Rebek et al. realised,[14] that these tetra-urea derivatives are self-complementary and exist in apolar solvents exclusively as hydrogen bonded dimers held together by a belt of intermolecular hydrogen bonds as shown in Fig. 9.4. This directionality of the C=O groups makes the single calixarene moiety chiral (C_4-symmetry) while the whole dimer is a meso-form with S_8-symmetry (see below). Enantiotopic protons or groups in either the ether residues (Y = CH$_2$Ph, CH$_2$COOEt) or the urea groups (R = CMe$_2$CH$_2$CMe$_3$) become diastereotopic in the dimer which provides further NMR evidence.

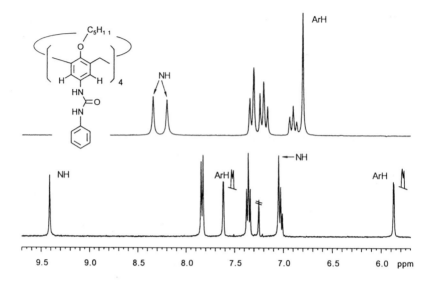

Figure 9.3. Section of the ^{1}H NMR-spectra of a tetra-urea **11a** (Y = C$_5$H$_{11}$, R = C$_6$H$_5$) in DMSO-d$_6$ (above) and CDCl$_3$ (below).

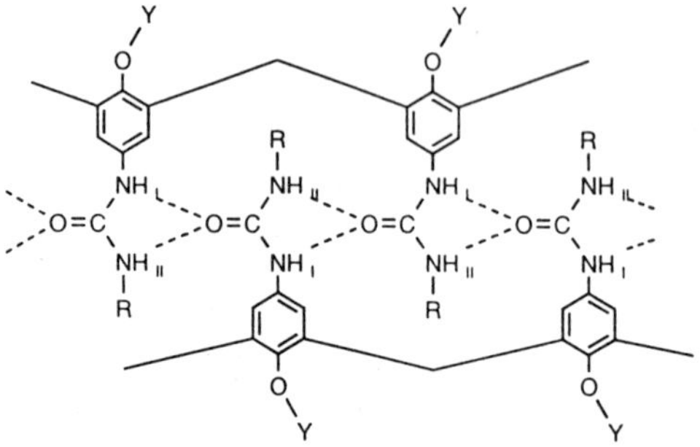

Figure 9.4. Schematic representation of hydrogen bonding in dimers of **11**.

This dimerization of tetra-ureas **11** is a rather general phenomenon with respect to structural modifications concerning the residues Y (various aliphatic residues including methyl) and R (aliphatic as well as aromatic residues). However, the conformation of tetramethoxy calix[4]arenes is not controlled by the dimerization.[15] Their tetra-urea derivatives exist already in the cone conformation (in contrast to the tetra-nitro precursor which prefers the partial cone conformation) under conditions, where only the monomeric form is present.

The formation of hetero-dimers $11_A·11_B$ in addition to the homo-dimers $(11_A)_2$ and $(11_B)_2$ in a mixture of two different tetra- ureas (11_A, 11_B) was considered as an additional proof of the dimerization.[16]

The structure proposed by Rebek was subsequently entirely confirmed for one example (**11b**, R = $C_6H_4CH_3$, Y = CH_2COOEt) by single crystal X-ray analysis.[17] (Fig. 9.5.) which is also in reasonable quantitative agreement (distances, angles) with the energy minimised structure. Remarkably, the NH···O=C distances are quite different, 2.85/2.84 Å for NH_{II} attached to the tolyl residue and 3.16/3.10 Å for NH_I attached to the calixarene skeleton, indicating hydrogen bonds of quite different strengths. In combination with shielding/deshielding effects this explains the strong splitting of the NH singlets observed in solution.

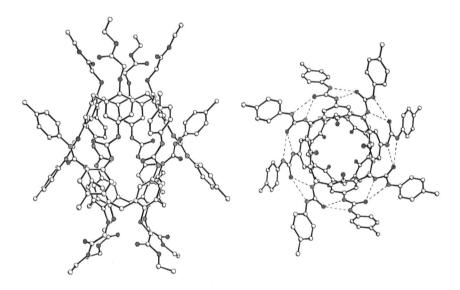

Figure 9.5. Single crystal X-ray structure of **11b** (R = $C_6H_4CH_3$, Y = CH_2COOEt). Disordered benzene molecules in the cavity, between the ester groups, and on other places in the crystal lattice are not shown.

Dimerization of tetra-ureas has been checked also by mass spectroscopy but the results were not convincing before neutral guests were replaced by cationic guests (see below).

9.2.2.2. Encapsulation of Guest Molecules

The dimerization seems to be strictly dependent on the "availability" of a suitable guest molecule, often the solvent (e.g. chloroform, benzene), to be included into the cavity formed in the dimer. This inclusion of a guest molecule can be deduced NMR-spectroscopically in different ways. For non deuterated guests upfield shifted signals ($\Delta\delta \approx 2.5\text{-}3.0$ ppm) are observed in addition to the usual signal(s) of the free guest.[14] In a mixture of two dimerizing ureas 11_A and 11_B three such signals may be observed, in agreement with its inclusion into dimers $(11_A)_2$, $11_A \cdot 11_B$, and $(11_B)_2$. The presence of two (or more) competitive guests (see below) in a solution of a single urea leads on the other hand to two (or more) different sets of signals for the two (or more) complexes with the same homo-dimer.

An independent experimental indication was obtained from the diffusion coefficients easily derived from pulsed gradient spin echo NMR-studies. For signals attributed to the encapsulated guest lower values, identical to those of the host, are found, than for the free guest.[18] E.g. for benzene the peak at 7.15 ppm leads to $2.08 \cdot 10^{-5}$ cm^2 s^{-1}, the peak at 4.4 ppm and several peaks for the tetra-urea calixarene lead to $0.034 \cdot 10^{-5}$ cm^2 s^{-1}.

Using cationic guests (e.g. tetraalkyl ammonium ions) it became possible to prove not only the dimerization of tetra-ureas, but also the guest inclusion by ESI MS.[19] Collision experiments make this inclusion highly probable as compared to complexes where the cation is loosely attached to the outer surface by cation-π interactions. It was even possible to demonstrate the size selectivity of the encapsulation in a qualitative way by competition experiments with cations of different size.

The only crystal structure available so far (see above) revealed the inclusion of benzene in the cavity formed by the dimerization. However, the included guest is completely disordered, obviously since the volume of this cavity ($V \approx 200$ Å^3) is too large to enforce a fixed position.

9.2.2.3. Thermodynamic Aspects

Up to date nothing is known about the stability of the dimers in solution, characterised for instance by equilibrium constants for reactions like

$$2\ 11 \rightleftharpoons (11)_2 \quad \text{or} \quad 2\ 11 + G \rightleftharpoons 11 \cdot G \cdot 11,$$

since no conditions have been found under which dimers and the single tetra-urea molecules can be observed simultaneously together. Heating (up to 110°C in C$_2$D$_2$Cl$_4$) usually does not change the NMR-spectra under usual concentrations (≈ 20 mM). Upon dilution finally (c < 1 mM) the dimers disappear, but they are replaced by various species partly associated via hydrogen bonds. The addition of dialkyl or diaryl ureas or of small amounts of hydrogen bond breaking solvents also leads to a destruction of the dimers but not to a simple equilibrium. From PGSE NMR studies it can be estimated, that roughly four molecules of DMSO are bound per molecule of monomeric **11** in benzene.[18]

Using a solvent which is less favoured as a guest (e.g. *p*-xylene) competitive complexation may be used to establish a relative affinity for different guests. Typical values are shown in Table 3.7. Own experiments gave a ratio of 750:600:25:1 for 1,2-difluorobenzene, benzene, hexafluorobenzene and *p*-xylene.[18]

Although there are some remarkable differences between different guests (in terms of guest inclusion this means "selectivities") they are by far not so pronounced as those observed for the hydrogen bond driven dimerization of a rigid cavitand (see below) leading to capsules with a similar shape. In addition to small differences in the internal volume this may be due to the more flexible conformation of the calixarene skeleton itself, as well as to the more flexible connection via the "urea-hinges".

In a mixture of two tetra-ureas **11$_A$** and **11$_B$** (differing for instance by R or Y) the three dimers (**11$_A$**)$_2$, **11$_A$·11$_B$**, and (**11$_B$**)$_2$ are formed in a more or less statistic ratio of 1:2:1, regardless whether R is aliphatic or aromatic.[16] There is, however, one important exception. When tosyl substituted tetra-ureas (R = SO$_2$-C$_6$H$_4$-CH$_3$) are combined with aryl substituted ones (e.g. R = C$_6$H$_4$-CH$_3$) only hetero-dimers are formed[20] (see below), (while the combination of tosyl with aliphatic residues leads to less than 10% hetero-dimer). Although the exact reason for this serendipitous observation is not really known, a favourable combination of NH acidity and O=C basicity may be assumed.

9.2.2.4. Kinetic Aspects

A half-life time of about 8 min was reported for the encapsulation of benzene (B) in the dimer (**11c**)$_2$ (Y = CH$_2$C$_6$H$_5$, R = C$_6$H$_4$F) in *p*-xylene (X) solution,[21] a process which (most probably) must be described at least as a bimolecular reaction

$$11c \cdot X \cdot 11c + B \rightleftharpoons 11c \cdot B \cdot 11c + X.$$

Therefore a half-life time is insufficient to characterise the reaction rate, unless the concentrations of the reactants are known.

A more detailed information about the kinetic stability of the dimers and about the rate of guest exchange was obtained by the tetra-urea **12**.[22] Since this molecule is C_{2v}-symmetric by its constitution it becomes C_2-symmetric by the direction of the carbonyl groups. Again two enantiomeric molecules are combined in the dimer, but due to the lower symmetry this does not result in a meso-form. The C_2-axis remains the only symmetry element of the dimer and the two calixarene parts are non-equivalent. Thus, for the aromatic protons of each of the two different phenolic units four different positions exist (Fig. 9.6.).

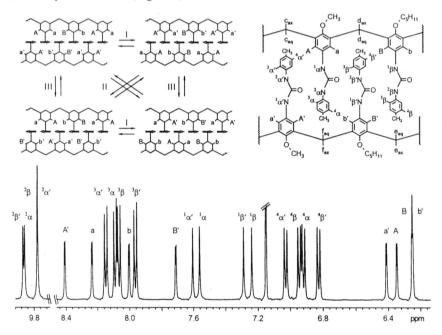

Figure 9.6. Section of the ^{1}H NMR-spectrum (500 MHz) of **12** in benzene-d_6 (bottom) and schematic representation of the possible homomerization pathways (top). The schematic formula (top) explains the peak assignment (based on NOEs/ROEs).

This allows to study three independent exchange reactions (I-III) by NMR (a fourth one leading to the identical situation cannot be detected).[a] Two of them (II, III) are possible only by dissociation/recombination while the third one, the change of the direction of the C=O functions could occur also within the dimer. An identical

---

[a] Two positions for Ar-H exist in the usual S_8-symmetrical dimer and consequently only one exchange reaction between these two positions can be studied.

rate for all three processes ($k_{I-III} = 0.055 – 0.069$ s^{-1}) suggests the same mechanism, and leads to an overall first order rate constant for the dissociation/recombination process of $k_{d/r} = 0.26 \pm 0.06$ s^{-1} (in benzene at room temperature). This is in reasonable agreement with the rate constant obtained for the exchange between free and encapsulated benzene $k_e = 0.47 \pm 0.1$ s^{-1} (determined by NOESY in a mixture of 77% C_6H_6 and 23% C_6D_6).[22]

Some more or less qualitative informations are also available for the mobility of the included guests within the capsule:

a) Fluorobenzene, for instance, shows large downfield shifts for the *o*- and *m*-protons (less for the *p*-proton) indicating an orientation with the fluorine and its *p*-position pointing towards the equator (the urea belt) of the capsule, while the *o*- and *m*-protons point towards the π-electrons of the cavity.[21] The same orientation is assumed for pyrazine, like in dimers of hydroxy cavitands (or carcerands) (see below) but in contrast, nothing is known about the barrier for its rotation.

Considering the different bridges between the *p*-positions of a symmetrical (C_{4v}-shaped) [1$_4$]metacyclophane, HN-C(O)-NH, O-H···O$^{(-)}$, and O-CH$_2$-O the internal volume of the cavities created should be rather similar (compare Table 9.2.), but subtle differences (sterically, electronically, flexibility) might have some crucial influence here.

b) The benzene molecule included in the cavity of (11b)$_2$ in the crystalline state is obviously completely disordered as revealed by the X-ray structure (see above).[17]

c) Two sets of signals with different intensities are seen for tricyclene as a guest, explainable by two different orientations of the included molecule with unequal populations which do not interconvert on the NMR-time scale.[20]

d) In contrast to other capsules, the barrier for cyclohexane encapsulated in the dimer (11d)$_2$ (Y = CH$_3$, R = C$_6$H$_4$CH$_3$) in toluene-d$_8$, determined from the coalescence of the signal of cyclohexane-d$_{11}$ (C$_6$HD$_{11}$), does not significantly differ from that of free cyclohexane ($\Delta G^{\neq} = 10.24/10.27 \pm 0.05$ kcal/mol).[23]

9.2.2.5. Stereochemistry, Chirality

The stereochemical properties of tetra-urea dimers are intriguing and challenging as well. As already mentioned for the simplest case, the directionality of the C=O groups makes each single calix[4]arene in the dimer chiral (C_4-symmetry), while the dimer as a whole is an achiral meso-form (S_8-symmetry) composed of a pair of enantiomers.

Inclusion of a chiral guest (like (1R)-(+)-nopinone) leads to a doubling of the NMR-signals for the calixarene since, two diastereomeric host-guest pairs R-P and

R-M[b] exist now within a dimer.[20] This holds as long as the direction of the C=O groups is fixed on the NMR-time scale, even if the guest is rapidly tumbling inside the cavity. It should be even true for a racemic guest mixture, where in addition S-P and S-M pairs would be present, the mirror images of the above mentioned species.

For a hetero-dimer formed by two different C_{4v}-symmetrical tetra-ureas, the directionality of the carbonyl groups leads to an overall C_4-symmetrical capsule I (Fig. 9.7.), existing as a pair of enantiomers (P1·M2 and M1·P2). Inclusion of a chiral guest should lead to the same doubling of the NMR-signals (for *both* tetra-urea parts) as before. However, the diastereomeric host-guest pairs (R-P1/R-M1 and R-P2/R-M2) exist now in different (enantiomeric) capsules.

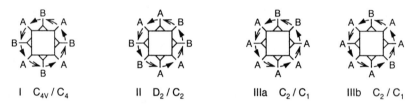

Figure 9.7. Schematic representation of the stereochemical properties of hetero- and homo-capsules. The symmetry classes are indicated for the whole dimer without/with directionality of the C=O groups (arrows).

Dimerization of C_{2v}-symmetrical tetra-urea derivatives (ABAB) such as **12** leads to an overall C_2-symmetry, and hence to chiral capsules **II** (see above). Although both calix[4]arene parts form a pair of enantiomers (due to the directionality of the C=O groups), they are diastereomers due to their combination within the capsule, giving two sets of signals in the NMR-spectrum.[22] The inclusion of a chiral guest should lead to a further splitting of signals.

The situation is even more complicated for a C_s-symmetrical derivative (AAAB), e.g. a mixed tetraether with one residue Y^1 and three residues Y^2. Two diastereomeric dimers are possible here, with a proximal (**IIIa**) or distal (**IIIb**) arrangement of the phenolic unit B.[24] Both are asymmetric and consequently chiral.

It is important to note, that in the hetero-dimer **I** the chirality is due to the directionality of the C=O groups. In homo-dimers **II** and **III** it stems from the mutual arrangement of the both calixarenes. The change of the C=O direction creates the opposite enantiomer for **I** but leads to the same enantiomer for the homo-dimers **II** and **III**.

---

[b] The chirality of the calix[4]arene parts is indicated by P/M while R/S is used for the included guest.

If additionally chiral centres (asymmetric C-atoms) are incorporated into the ether or urea residues (e.g. R = -(S)-CH(C_4H_9)-C(O)-O-CH_3) two diastereomeric forms of this single tetra-urea (SP, SM) exist in a dimer due to the orientation of the carbonyl groups (P, M), and consequently for a hetero-dimer formed with a "normal" tetra-urea two diastereomeric forms (SP1·M2, SM1·P2) should exist. This has not been observed so far. Only diastereomeric complexes with racemic nopinone were reported.[20]

Finally, this additional chiral centre could induce a certain direction of the C=O groups (e.g. SP being more stable than SM, and consequently RM more stable than RP). Then the formation of hetero-dimers SP·RM should be preferred in a racemic mixture of tetra-ureas R and S over the homo-dimers SP·SM and RP·RM. Again this has not yet been observed.

9.2.2.6. Larger assemblies

Two tetra-urea derivatives may be connected by an (additional) covalent linker to a double calixarene. The dimerization of the urea parts then may lead to either linear polymeric assemblies, or to cyclic oligomeric structures, in the extreme case just to an intramolecularly self-assembled capsule. Both extreme possibilities have been realised by Rebek et al.[24,25,26]

Linear polymeric assemblies are obtained by bis-(tetra-ureas) of type **13** or **14** in which two tetra-urea subunits are linked via a single chain between the ether functions in a more or less flexible way. The synthesis of these compounds is straight forward. O-alkylation of t-butylcalix[4]arene leads to the *syn*-triether which is further alkylated by ethyl bromoacetate. Ipso-nitration, reduction and reaction with an isocyanate give the tetra-urea derivative. Finally, hydrolysis of the ester group and coupling with a diamine leads to various molecules **13**. (A mono-protected diamine may be used, to obtain a double calixarene **14** with two different tetra-urea subunits via coupling, deprotection and coupling with a second tetra-urea).[24,25]

The formation of linear "polycaps" is deduced for instance by a low field shift of the NH_{II}-protons. The spectra are broad and unresolved, since for each simple capsule exist already two diastereomeric situations (with 8 different NH_{II}-protons), with the amide functions in a proximal or distal arrangement (see types **IIIa/IIIb** discussed above). Polymer formation may be induced by the addition of a suitable guest, while the gradual addition of a simple, non polymeric dimer leads to a degradation of the polymeric associates due to the formation of hetero-dimeric structures:

$$-[13]_n- + n\,(11)_2 \rightleftharpoons n\,(11\cdot13\cdot11)$$

13 R = R', Y = Y'
14 R ≠ R' and / or Y ≠ Y'

Various discrete assemblies can be formed, using the preferred hetero-dimerization of tetra-aryl- and tetra-tosyl-ureas. By mixing the "complementary" tetra-urea moieties in form of simple (**11**), double (**13**) or triple calixarene derivatives (**15**) the dimer **D1**, the bis-dimer **D2** and the tris-dimer **D3** are found (as the only species) by NMR in CDCl$_3$ (Fig. 9.8.). The polycaps **P1** and **P2** consisting only of hetero-dimeric structures are formed from a mixture of the two homo-biscalixarenes **13** and the hetero-biscalixarene **14**, respectively.[24]

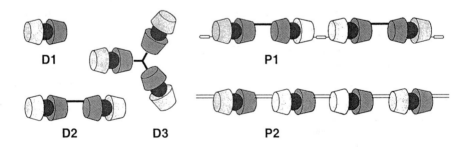

Figure 9.8. Schematic representation of discrete assemblies and "alternating" polymers using the exclusive formation of hetero-dimers. Tetra-urea structures with R = tosyl or tolyl are symbolised by a light and a dark cone.

Two tetra-urea calix[4]arenes are connected by a single hexamethylene chain between their urea functions in bis-(tetra-ureas) of type **16**.[26] The synthesis starts with the trinitro tetraether (by partial nitration) which is substituted by an amino group in the fourth *p*-position by iodination, substitution with phthalimide and hydrazinolysis. Coupling with 1,6-diisocyanatohexane gives the urea linked bis-calixarene. Hydrogenation of the nitro groups followed by treatment with excess of (*n*-heptyl)phenylisocyanate) finally leads to **16**, which shows the expected ^{1}H NMR-spectrum in DMSO-d_6 (e.g. five signals for the NH-protons adjacent to aryl groups in the expected ratio 1:2:2:1:1). A down field shift of the signals of the NH$_{II}$-protons in CDCl$_3$ indicates the formation of dimeric tetra-urea structures, for which three possibilities are considered, as indicated schematically in Fig. 9.9.

The formation of an unimolecular, clamp-like capsule **CD** as the predominant species is suggested (mainly) on the basis of ESI-MS spectral evidence,[19] using cationic guests like N-methyl quinuclidinium, while eight signals for the aryl NH$_{II}$-protons instead of the expected six signals indicate a more complicated situation. Addition of a homo-dimer (**11**)$_2$ (with R = tolyl) does not lead to the formation of additional species with hetero-dimeric structures which are formed, however, if a homo-dimer (**11**)$_2$ (with R = tosyl) is added (evidence by NMR and ESI MS). This underlines again the preference for hetero-dimers described above.

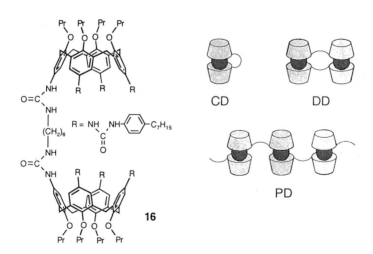

Figure 9.9. Possible species formed by dimerization of the tetra-urea moieties of **16**.

9.2.3. The Melamine-Barbiturate Motif

Melamines and their counter parts barbiturates or cyanurates are able to associate via three intermolecular hydrogen bonds between two complementary molecules forming hydrogen bonded polymeric and oligomeric structures (Fig. 9.10.) in the crystalline state.[27]

Whitesides et al. demonstrated that depending on the residues R well-defined hydrogen bonded hexamers ("rosettes") can be obtained not only in the crystalline state, but also in apolar solvents. Appropriate preorganization of the rosette components by covalent attachment to a molecular skeleton led to well defined hydrogen bonded associates differing by the number of particles and thus exhibiting different stability due to the entropic factor.[28]

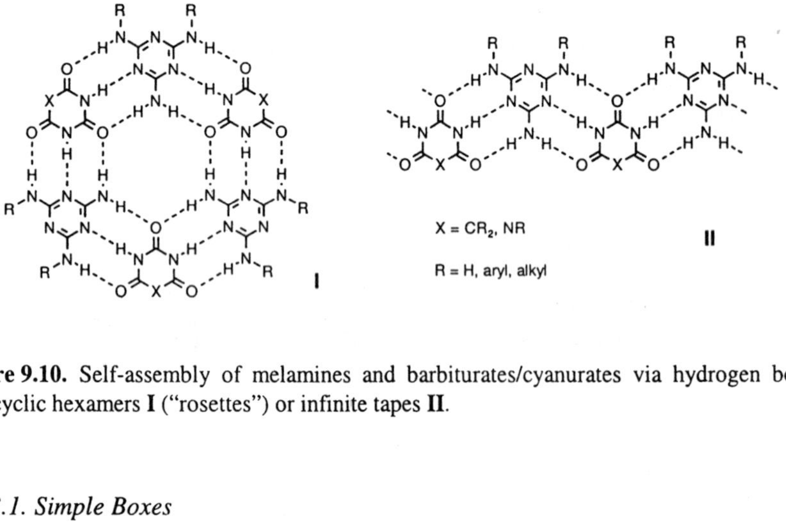

Figure 9.10. Self-assembly of melamines and barbiturates/cyanurates via hydrogen bonds into cyclic hexamers **I** ("rosettes") or infinite tapes **II**.

9.2.3.1. Simple Boxes

This structural motif melamine-barbiturate/cyanurate was used by Reinhoudt et al. for the noncovalent organisation of calixarenes via hydrogen bonds.[29] The bis-melamine derivatives **17** were prepared by reaction of 1,3-diamino calixarenes with cyanurchloride followed by stepwise substitutions with ammonia and an aliphatic amine. Interaction of compounds **17** with barbiturates **18** or cyanurates **19** results in the formation of well-defined, box-like assemblies consisting of nine hydrogen bonded particles (6 × **18** or **19**, 3 × **17**) connected by 36 hydrogen bonds.

219

Y= C₃H₇, C₁₂H₂₅

R¹= H, NO₂, NH₂, CN, NHC(O)CH₃

R²= C₄H₉, CH₂Ph, CH(CH₃)Ph

R= CH₂-CH₂-C(CH₃)₃, 4-C₆H₄-C(CH₃)₃, CH₂-CH₂-C(C₆H₅)₃

These aggregates are stable on the NMR time scale at ambient temperature in apolar solvents such as CDCl$_3$ or toluene, while the addition of small amounts of polar solvents such as DMSO or MeOH leads to their complete destruction.

Usually, the ^{1}H NMR-spectra of $(17)_3 \cdot (18)_6$ in CDCl$_3$ contain only two sharp singlets for the imide protons of the barbiturate at 13-14 ppm. This strong downfield shift is characteristic for the melamine-barbiturate hexamer while the simple pattern is in agreement only with an overall D$_3$-symmetry. In order to form such an aggregate the bis-melamine calix[4]arene has to adopt the C$_2$-symmetrical (staggered) conformation while the C$_s$-symmetrical (eclipsed) conformer can form either a C$_{3h}$-symmetrical or a C$_s$-symmetrical eclipsed assembly (Fig. 9.11.). The simultaneous formation of all three supramolecular diastereomers was detected by ^{1}H NMR-spectroscopy in the case of rosettes based on the sterically hindered cyanurates **19a-c** (R = CH$_2$-CH$_2$-C(CH$_3$)$_3$, C$_6$H$_4$-C(CH$_3$)$_3,$ CH$_2$-CH$_2$-C(C$_6$H$_5$)$_3$).[30]

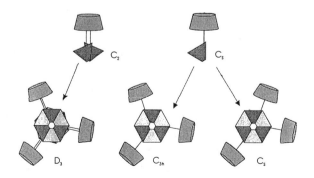

Figure 9.11. Three possible diastereomeric "boxes" and their symmetry.

An additional structural proof was gained from a single crystal X-ray analysis for one example of $(17a)_3·(18)_6$ ($17a$: $Y = C_3H_7$, $R^1 = NO_2$, $R^2 = C_4H_9$).[29] Indeed, in the crystalline state two tightly stacked rosettes are formed which are connected by three calixarene spacers in the supposed box-like arrangement (Fig. 9.12.). The calixarenes adopt a strongly pinched, C_2-symmetrical cone conformation in which the two aryl rings bearing the melamine fragments are nearly parallel. Due to this pinching no space is left for the inclusion of any guest. The anti-parallel orientation of the two rosette motifs makes the whole aggregate chiral (D_3-symmetry), in agreement with the NMR-data. However, in solution fast racemization may take place due to the reversibility of the hydrogen bonds.

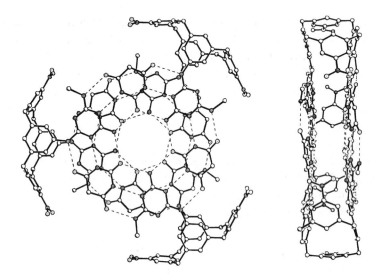

Figure 9.12. Single crystal X-ray structure of $(17a)_3·(18)_6$. ($17a$: $Y = C_3H_7$, $R^1 = NO_2$, $R^2 = C_4H_9$) Butyl and propyl chains of **17a** as well as ethyl groups of **18** are omitted for clarity.

The addition of 1.5 mol of silver triflate ($AgOOCCF_3$) per mole of the complex to the solution of $(17)_3·(18)_6$ or $(17)_3·(19)_6$ in $CDCl_3$ led to the detection of a 1:1 complex by MALDI-TOF mass spectrometry. However, the presence of Ag^+-complexing fragments in the bis-melamine (e.g. $R^1 = CN$ or $R^2 =$ benzyl) or/and cyanurate/barbiturate structure is crucial for the effectiveness of this method.[31]

The calixarene boxes are effective platforms for the preorganization of multiple functional groups. This can be achieved both by modifications of calixarene components and by using various barbiturates and cyanurates. Most of groups R[1] attached to the wider rim of calixarene bis-melamines do not disturb the formation of the rosettes. However, in the case of amide groups well defined assemblies were not formed due to the unfavourable competition of hydrogen bonding.

Mixing of solutions containing the two homo-aggregates $(17a)_3 \cdot (18)_6$ and $(17b)_3 \cdot (18)_6$ resulted in the additional formation of the two possible hetero-aggregates $(17a)_2 \cdot (17b) \cdot (18)_6$ and $(17a) \cdot (17b)_2 \cdot (18)_6$ in a statistical ratio (Fig. 9.13.). The complicated mixture could be well characterised by ^{1}H NMR-spectroscopy and MALDI-TOF mass spectrometry. Combination of N compounds **17** with a single compound **18** or **19** in the ratio 1:3 should lead to a mixture of $P = N + N(N-1) + [N(N-1)(N-2)]/6$ different boxes, nowadays called a "dynamic combinatorial library" of hydrogen-bonded assemblies.[32] The diversity increases in a yet unpredictable way by the use of different compounds **18/19**.

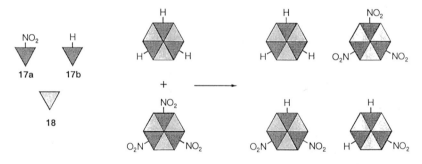

Figure 9.13. Combination of two "homo-aggregates" $(17a)_3 \cdot (18)_6$ and $(17b)_3 \cdot (18)_6$.

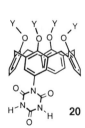

If a calixarene bis-melamine, e.g. **17c** (Y = $C_{12}H_{25}$, R[1] = H, R[2] = C_4H_9) is combined with the calixarene cyanurate **20** (instead of a simple barbiturate **18** or cyanurate **19**) the assembly consists of 9 calixarenes. Three of them (**17c**) form the central box while the other six (**20**) create big clefts below and above it.[33]

The box-like assemblies described so far usually have D_3-symmetry and exist in form of two enantiomers with "supramolecular helicity" (M, P). If in addition two chiral centres are introduced in the residue R[2] (R, S) two diastereomeric assemblies might result (RR/M and RR/P and their enantiomers SS/P and SS/M). However, obviously only **one** of the two possible diastereomers is formed if **17d,e**

(R^2 = (R)- or (S)-CH(CH$_3$)C$_6$H$_5$) is combined with two moles of **18**, since the ^{1}H NMR-spectra show the typical pattern for the D$_3$-symmetrical aggregates and are similar to those of the nonchiral analogues. This formation of a single diastereomer is further corroborated by a much stronger circular dichroism in contrast to the individual calixarene bis-melamines. In the case of the "mixed" (RS) bis-melamine the NMR-spectrum of the aggregates showed only broad, unresolved peaks, demonstrating the strong influence of the peripheral chirality on the stability of the helical hydrogen bonded box. Obviously the chiral residues R^2 require a certain orientation within a stable rosette, which could be achieved for the RS isomer only in a C$_s$-symmetrical conformation leading to the less favourable C$_{3h}$-symmetrical box. From the ROESY spectra the authors conclude, that the RR-isomer induces the formation of the M-form and vice versa.

9.2.3.2. Larger Aggregates

Based on the melamine-barbiturate box as module, Reinhoudt et al. developed the next generation of well defined nanostructures.[34] The double calixarene **21** (Fig. 9.14.) connected via two of the four melamine fragments was prepared starting with a mono-BOC protected diamino calixarene from which a monomelamine derivative was obtained in the usual way. After deprotection, the second melamine was built up

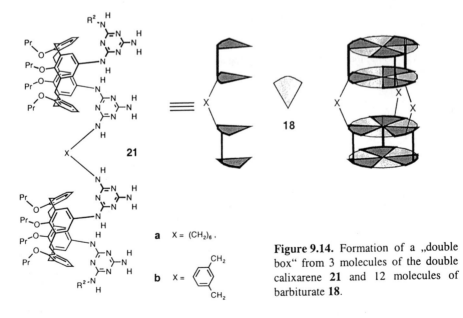

Figure 9.14. Formation of a „double box" from 3 molecules of the double calixarene **21** and 12 molecules of barbiturate **18**.

analogously using a mono-BOC protected diamine in the last step. Deprotection and reaction with the monochlorotriazine precursor gave **21**. (Similar compounds consisting of two different calixarenes would be available in the same way.)

Interaction of **21** with four moles of **18** leads to the aggregate $(21)_3 \cdot (18)_{12}$ consisting of 15 particles held together by 72 hydrogen bonds. The formation of $(21)_3 \cdot (18)_{12}$ was proved both by ^{1}H NMR-spectroscopy and MALDI-TOF mass spectrometry with Ag^+ assisted ionisation. Competition experiments between **21** and the bis-melamine **17** showed that the latter forms more stable aggregates $(17)_3 \cdot (18)_6$. Incomplete filling of the space between two double rosettes was assumed as a possible reason for the lower stability of $(21)_3 \cdot (18)_{12}$, however one can suppose that also the entropic factor plays a considerable role in this difference. Mixing of $(17)_3 \cdot (18)_6$ and $(21)_3 \cdot (18)_{12}$ does not lead to the formation of mixed aggregates (such as $(21)_2 \cdot (17)_2 \cdot (18)_{12}$ or $21 \cdot (17)_4 \cdot (18)_{12}$), while hetero-association was observed in the case of $(21a)_3 \cdot (18)_{12}$ and $(21b)_3 \cdot (18)_{12}$. This provides a possibility to establish combinatorial libraries of dynamic nanostructures consisting of 15 particles.

The building principle described so far can be further extended to create polymeric hydrogen-bonded nanostructures. For this purpose calixarene bis-cyanurates **22** were prepared by reaction of calixarene diamines with nitro biuret in DMF and subsequent closure of the cyanurate ring (diethylcarbonate/NaOEt).[35] In contrast to simple barbiturates **18** and cyanurates **19** compounds **22** are not able to form well defined aggregates with **17** but only hydrogen bonded polymers (Fig. 9.15.).

While a mixture of **17a** and **22a** ($Y = C_3H_7$) resulted in the formation of a precipitate which was soluble only in DMSO, the more lipophilic compounds **17c** and **22b** ($Y = C_{12}H_{25}$) led to polymeric aggregates soluble also in apolar solvents. The ^{1}H NMR-spectrum of $[(17c)_3 \cdot (22b)_3]_n$ in $CDCl_3$ contains two characteristic singlets for the imide protons of cyanurate (14-15 ppm) with line shapes typical for polymeric structures. The self-assembly of **17c** and **22b** on an apolar graphite surface resulted in the formation of regular rod-like patterns which were characterised by Tapping-Mode Scanning Force Microscopy (TM-SFM). A good correlation was found between the experimentally determined and the calculated dimensions of the nanostructures obtained. The future use of these rod-like assemblies as a basis to construct molecular wires is envisaged.

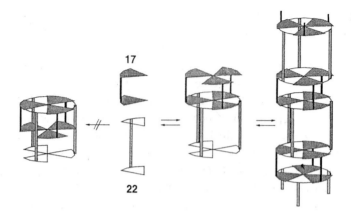

Figure 9.15. Formation of polymeric aggregates from calix[4]arene derivatives **17** and **22** using the melamine-cyanurate motif.

9.2.4. Further Dimerizations

Recently, the synthesis of a calix[4]arene derivative fixed in the 1,3-alternate conformation was described in which two opposite phenolic units are substituted in their *p*-positions by 2-ureidopyrimidine-4(1H)-one groups.[36] These systems dimerize in a more or less rigid, planar arrangement held together by four hydrogen bonds in an AADD sequence (A = acceptor, D = donor).

The dimer formed by the calixarene thus is connected by eight hydrogen bonds, leading to an association constant of $K_{ass} > 10^6$ M^{-1} in toluene. The dissociation may be achieved by competitive hydrogen bonding with DMSO (>50%) or triflate anions.

Such a connection via rigid, planar, hydrogen bonded segments is probably also the best way to create larger closed cavities via the self-assembly of concave building blocks.

9.3. Resorcarenes

9.3.1. The Parent Octahydroxy Tetramers

Already Cram[37] has shown that rccc resorcarenes **24** adopt a more or less regular crown (cone) conformation in the crystalline state which is stabilised by four intramolecular hydrogen bonds between neighbouring hydroxyl groups (Fig. 9.16.). The four remaining hydroxyls are available for *inter*molecular hydrogen bonding. This property of resorcarenes allows the formation of various hydrogen bonded architectures in the crystalline state.

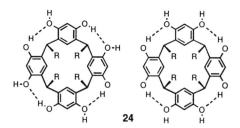

Figure 9.16. Two possible arrangements of hydrogen bonds in resorcarene octols **24**: chiral C_4-symmetrical (left) and achiral C_{2v}-symmetrical (meso-form, right).

Compound **24a** (R = Me) forms a crystalline solvate with pyridine in which four hydrogen-bonded pyridine molecules, arranged in a C_{2v}-symmetrical manner (Fig. 9.16.), enlarge the cavity of the resorcarene, while one disordered pyridine molecule plays the role of a guest[38] (Fig. 9.17). Crystallisation of **24b** from pyridine or γ-picoline in the presence of CH_3NO_2, CH_2Cl_2, CH_3CN gave similar structures showing an inclusion of the corresponding molecule as a guest in the extended cavity.[39]

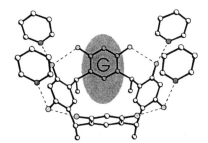

Figure 9.17. Inclusion of guests in the extended self-assembling cavity of resorcarene **24a** (R = Me). G = pyridine, CH_3NO_2, CH_2Cl_2, MeCN.

Various hydrogen bonded polymeric structures are obtained upon co-crystallisation of **24** with bispyridines and pyrimidine[40] demonstrating that resorcarenes **24** are promising tectones for crystal engineering. The opposite extreme are crystalline structures in which the single resorcarene molecules (rctt-isomers) are completely separated by (hydrogen bonded) solvent molecules (quasi-solutions).[41]

Resorcarenes **24** form also hydrogen bonded assemblies with a closed cavity in the crystalline state. Two principal types of such structures were reported, namely carcerand-like dimers and octahedral hexamers.[42]

Co-crystallisation of **24b** ($R = CH_2CH_2C_6H_5$) and C_{60} from *o*-dichlorobenzene/2-propanol (probably studied to entrap C_{60} in a closed resorcarene assembly) resulted in the formation of hydrogen bonded resorcarene dimers, while disordered C_{60} molecules fill the voids in the crystal lattice.[43] Two concave, C_{4v}-symmetrical resorcarene molecules, related by a mirror plane (Fig. 9.18.a) are coupled together via four hydrogen bonded 2-propanol molecules forming a closed spherical cavity of 230 Å^3, which contains very disordered solvent molecules.

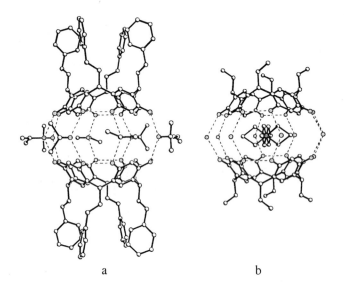

Figure 9.18. Crystal structure of $(24b)_2 \cdot 4i$-PrOH (a) and $(24c)_2^- \cdot 8H_2O \cdot Et_4N^+$ (b).

A similar structure was reported for $(24c)_2^- \cdot 8H_2O \cdot Et_4N^+$ obtained by co-crystallisation of **24c** (R = Et) with Et_4NClO_4 from EtOH.[44] In this case the negatively charged dimeric capsule is formed by two resorcarene molecules connected by eight hydrogen-bonded water bridges (Fig. 9.17.b). One partly disordered Et_4N^+ is included in the cavity so that the nitrogen atom is positioned at a crystallographic centre of symmetry. The negative charge is delocalized over the oxygen atoms of resorcinol rings and water molecules. The distances between the carbon atoms of Et_4N^+ and resorcinol rings are in accordance with C-H---π host-guest interactions.

The crystal structures of $(24b)_2 \cdot 4i$-PrOH and $(24c)_2 \cdot 8H_2O \cdot Et_4N^+$ demonstrate the ability of resorcarenes to employ several hydrogen bonded solvent molecules for the creation of self-assembled molecular containers.

Resorcarene **24d** ($R = C_{11}H_{23}$) solubilizes β-methyl glucopyranoside (GP) in CCl_4 and $CDCl_3$.[45] Based on a very large upfield shifts of the ^1H NMR signals of the OCH_3 groups ($\Delta\delta$ = -3.58 ppm) and ^1H-^{13}C NOE experiments the formation of a highly lipophilic complex $(24d)_2 \cdot GP$ was postulated where the GP-molecule is included in the π-basic cavity of a resorcarene dimer. The exact structure of the dimeric capsule and the reason for the pronounced selectivity for the β- over the α-anomer could not be deduced from the NMR studies. However, it seems reasonable to assume a dimeric structure resembling those shown in Fig. 9.18.b.

McGillivray and Atwood[46] found conditions under which six molecules of **24a** and 8 water molecules arrange in the crystalline state in an octahedral-cubic assembly held together by 60 hydrogen bonds (Fig. 9.19.). The volume of the cavity thus formed (V = 1375 Å^3) would be big enough to include guests such as porphyrins, fullerenes etc. Electron density maxima were found within the interior of $24a \cdot 8H_2O$, however, guest species could not be identified from the X-ray experiment.

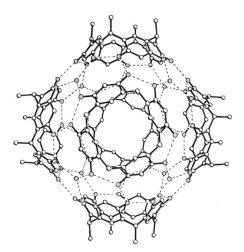

Figure 9.19. Crystal structure of $(24a)_6 \cdot 8H_2O$. Hydrogen atoms are omitted, hydrogen bonds are indicated by dotted lines between oxygens.

Although the hydrogen atoms of hydroxyl groups could not be determined experimentally, it was postulated that in order to reach the optimal saturation of hydrogen bond donors and acceptors four resorcarene molecules should have a C_4-symmetrical arrangement of the hydrogen bonds (chiral sites) while the two remaining ones should be C_{2v}-symmetrical (achiral sites). Under this assumption two types of contacts between three resorcarene molecules are possible in which the water molecule forms two hydrogen bonds as a donor and one as an acceptor and vice versa.

Based on previous molecular weight determinations of Aoyama et al. it was concluded that a hexameric assembly exists also in benzene solution. The ^{1}H NMR-spectrum of **24d** in benzene-d_6 is rather broad and complicated. It contains two well-defined signals for the protons of methine bridges in a 2:1 ratio which would agree with the C_4- and C_{2v}-symmetrical resorcarene molecules forming the hexameric assembly.

However, the type of hydrogen bonded aggregation of **24d** strongly depends on the nature of the solvent. In CDCl$_3$ only one signal for the methine protons of the bridges is found for **24d** and a down field shifted signal of water molecules hydrogen bonded to the phenolic hydroxyl groups. Although it was postulated that an "opened" hydrogen bonded complex **24d**·4H$_2$O with water is formed,[47] a dimeric structure like in **(24c)$_2$**·Et$_4$N$^+$·8H$_2$O (Fig. 9.18.b) cannot be entirely ruled out.

A similar arrangement of six molecules was observed for the pyrogallol derived tetramer **25a** (R = i-Bu) in the crystalline state (Fig. 9.20.). The centro-symmetrical octahedral assembly is held together by 72 hydrogen bonds between the 72 phenolic hydroxyl groups.[48] In contrast to resorcarene **24a** the molecules of **25a** do not require hydrogen bonding solvent bridges to form the hexameric aggregate. To reach the mutual saturation of all hydrogen bond donors and acceptors a C_4-symmetrical chiral arrangement of hydrogen bonds must be assumed. An "intramolecular" cavity of at least 1520 Å^3 is estimated for the internal volume of the assembly. Ten acetonitrile molecules (solvent of crystallisation) could be found within the interior of **(25a)$_6$**. Six of them occupy the cavities of six hydroxyresorcarene molecules while the for remaining ones fill the rest of the super-cavity.

Theoretically, the hexamer **(25a)$_6$** should be thermodynamically more stable than the analogous **(24a)$_6$**·8H$_2$O since it consists of a smaller number of particles (entropic factor) held together by a larger number of hydrogen bonds (enthalpic factor). However, no convincing evidence for the existence of **(25a)$_6$** in solution was reported so far.

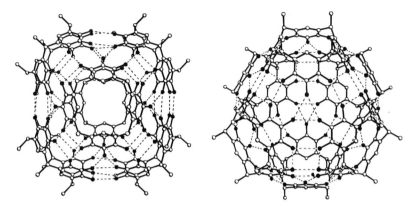

Figure 9.20. Crystal structure of (25a)$_6$. The *i*-propyl fragments of the *i*-butyl groups are omitted for clarity.

9.3.2. Hydrogen Bonded Dimers of Hydroxyl-Cavitands

Like calix[4]arenes fixed in the cone conformation, the all-cis resorcarenes still have a certain degree of conformational flexibility and may assume a C_{2v}-symmetrical boat conformation which is less favourable for their dimerization. Connection of hydroxyl groups of adjacent resorcinol units leads to the more rigid (C_{4v}-symmetrical) cavitands, which due to their rigidity should be more suitable for all kinds of self-assembled structures.

Tetrahydroxy cavitands **26** have been frequently used as building blocks for the synthesis of covalently linked carcerands **27**, using various structural elements -X- to connect two molecules of **26** via the four hydroxyl groups.[49,50]

Even simpler is the self-assembly of its dianion, formed in the presence of a base (usually DBU = 1,8-diazabicyclo[5.4.0]undec-7-ene) and a suitable guest.[51,52] This assembly is held together by four strong charged hydrogen bonds O-H···O$^{(-)}$ and is destroyed upon acidification. Thus, encapsulation and release of guests may be switched by appropriate adjustment of pH (Fig. 9.21.).

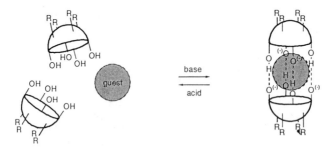

Figure 9.21. Schematic representation of acid/base switched guest encapsulation and release by hydroxycavitands **26**.

Competitive complexation experiments with different guests (in $CDCl_3$ or nitrobenzene-d_5) led to a sequence of relative complexation constants (Table 9.1.) which is more or less analogous to the template ratio established for the formation of the corresponding O-CH_2-O bridged carcerands **27** (X = CH_2).[53]

Table 9.1. Relative complexation constants (normalised to benzene) for the inclusion of various guests in the hydrogen bonded dimer $(26)_2^{4-}$. Available results for tetra-urea dimers $(11)_2$ are given for comparison.

Guest	K_{rel} $(26 \cdot G \cdot 26)^{4-}$		$11 \cdot G \cdot 11$
	in nitrobenzene-d_5	in $CDCl_3$	in p-xylene-d_{10}
pyrazine	1800	1300	3.2
methyl acetate	780	--	
dioxane	450	77	
DMSO	110	14	
pyridine	13	9.6	1.2
acetone-d_6	2.4	0.9	
benzene-d_6	1.00	1.00	1.00

This analogy is further established by studies with a series of partially hydroxylated cavitands (bowls **28-31**) and bis-cavitands (bis-bowls **32**, **33**), covalently connected by one or two -O-CH_2-O- bridges.[54] While the monohydroxy bowl **28** forms no complex, dimerization with guest inclusion (pyrazine) is observed for the two diols **29, 30** and the triol **31** upon addition of base (DBU). The doubly linked bis-bowl **33** forms a complex with and without DBU, while the mono-linked bis-bowl **32** requires the addition of DBU for complexation. Thus, at least two bridges, either covalent links (-O-CH$_2$-O-) or charged hydrogen bonds (-O$^{(-)}$···H-O-) are necessary to obtain a stable dimeric complex, a statement which is confirmed by various further compounds in which the OH groups are replaced by H or OCH$_3$.

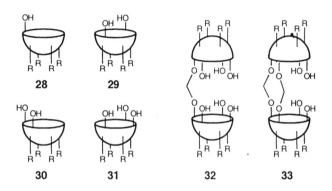

Base induced guest inclusion occurs also with the macrocyclic tetracavitand **34** as indicated in Fig. 9.22.[55] Again this formation of a "double container" via O-H···O$^{(-)}$ bridges parallels the corresponding "covalent encapsulation" via O-CH$_2$-O bridges.[56]

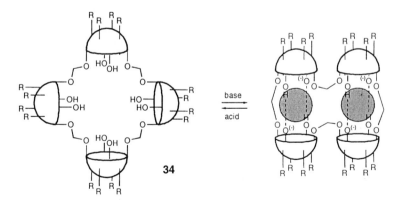

Figure 9.22. Schematic representation of base induced guest inclusion in a hydrogen bonded "double container" (or guest induced intramolecular self-assembly).

If a pyrazine molecule is included in a hetero-dimer consisting of two cavitands differing by their residues R, it shows two different ^{1}H NMR signals (Fig. 9.23.). Their *meta*-coupling proves an orientation (I) where the nitrogen atoms point towards the equator (later confirmed by an X-ray structure, see Fig. 9.25), while the orientation II with the nitrogen atoms pointing to the poles would result in *ortho*-coupling. From the coalescence temperature (T_c = 353 K, $\Delta\delta$ = 14,3 Hz) an energy barrier of $\Delta G^{\neq}$ = 18,3 kcal/mol was calculated for the rotation around the pseudo-C_2-axis.

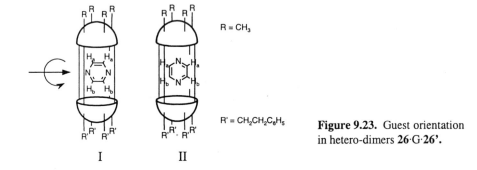

Figure 9.23. Guest orientation in hetero-dimers **26·G·26'**.

An even more detailed information about the dynamic stereochemistry for complexes of type $(26 \cdot G \cdot 26)^{4-}$ (and analogous hemicarceplexes and carceplexes) was recently obtained by Chapman and Sherman from variable temperature NMR.[57] All X-ray structures of such assemblies (or carceplexes) reveal, that the two bowls are twisted towards each other (rotation around the fourfold axis) by angles of 13-21° (compare Fig. 9.25.). Hence the time averaged D_{4h}-symmetry observed in solution is reduced to D_4 in the crystal and at sufficiently low temperature in solution. This chiral environment renders enantiotopic groups of an included guest diastereotopic (e.g. the methyl groups in DMSO). If the included guest is non-symmetric with respect to the C_2-axes (e.g. DMSO), the symmetry of the whole assembly is further reduced to C_4. Thus, protons of the two structurally identical bowls, which are enantiotopic with a symmetrical guest such as pyrazine, become heterotopic.

In fact an energy barrier of $\Delta G^{\neq}$ = 13.4 ± 0.2 kcal/mol was determined based on the coalescence of the two singlets for the methyl groups of DMSO and attributed to the interconversion of the two "twistomers".[57] A lower energy barrier of $\Delta G^{\neq}$ = 12.6 ± 0.2 kcal/mol derived from the coalescence of the inwards pointing protons of the O-CH_2-O-bridges of the cavitand was attributed to the rotation of DMSO around the C_2-axes, as schematically shown in Fig. 9.24. These values are very similar to barriers found in a corresponding carceplex with O-CH$_2$-O-bridges between the two bowls,[49] demonstrating again the "similarity" between this covalent linkage and the charged hydrogen bonds.

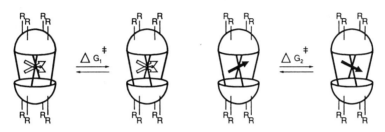

Figure 9.24. Schematic representation of conformational interconversions of inclusion complexes of type **26·G·26**. Left: Interconversion of "twistomers" with a still freely rotating guest. Right: Rotation of the guest (indicated by an arrow) in a frozen twistomer.

As already mentioned a single crystal X-ray structure could be obtained for the inclusion complex of **26a** (R = Me) with pyrazine,[52] confirming for the crystalline state all the observations made in solution. Fig. 9.25. shows the structure from two different directions. To compare the size of the cavity with similar complexes the distances of the molecular reference planes (methylene or methine carbons) and the diagonal distances of relevant atoms at the equator are collected in Table 9.2. These data, taken from the X-ray structures demonstrate, that in fact the tetra-urea dimers **(11)$_2$** have a somewhat larger internal volume.

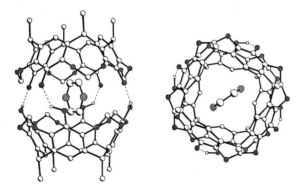

Figure 9.25. Single crystal X-ray structure of **(26a)$_2$**$^{4-}$ with included pyrazine. (DBU cations are not shown.)

Table 9.2. Comparison of the size of different dimers of calixarenes, resorcarenes and cavitands.

dimer	distance of mean planes (Å)	diagonal distance at the equator (Å)
(**11b**)$_2$	9.78	10.85 (carbonyl C)
		9.30 (calix N)
(**26a**)$_2^{4-}$	9.24	9.10
(**24b**)$_2 \cdot 4i$-PrOH	9.29	9.45
(**24c**)$_2 \cdot 8H_2O$	9.14	ˇ9.52

9.3.3. Extended Cavitands

Resorcarene based cavitands have also been used to create larger cavities, in the hope to include larger guests or more than one guest molecule. Compounds **35** easily available from the tetrabromo cavitand by Suzuki coupling with *p*-nitrophenyl boronic acid, reduction of the nitrogroups and acylation with various aliphatic acid chlorides. These "deep-cavity" resorcarenes (**35b** and **35c**) are self-complementary and dimerize in solvents such as toluene via N-H···O=C hydrogen bonding between amide functions of opposite monomers in a manner resembling somehow the tetra-urea calix[4]arenes discussed above.[58] The dimerization is concentration dependent (leading to K_D = 1700 ± 250 at 295 K for **35b** in toluene) and temperature dependent (ΔH = -21 ± 1.3 kcal mol^{-1}, ΔS = -58 ± 5 cal mol^{-1} K^{-1}).

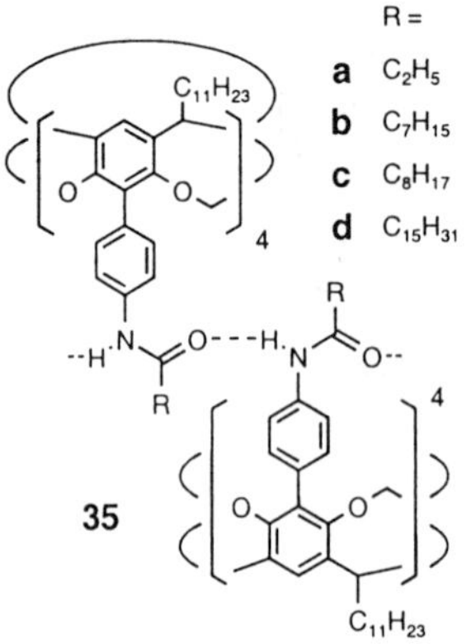

Although there should be an egg-shaped cavity with an estimated volume of 440 Å^3 suitable guests did not show any sign of inclusion. Obviously one aliphatic chain (heptyl or octyl) of each monomer is self-included, occupying 57% of the cavity. This self-inclusion seems even necessary for the dimerization.[59] While **35b** and **35c** form also a heterodimer neither **35a** nor **35d** dimerize in toluene. Presumably their alkyl chains are too small or too large to fill the cavity sufficiently.

R =
a C$_2$H$_5$
b C$_7$H$_{15}$
c C$_8$H$_{17}$
d C$_{15}$H$_{31}$

35

Reaction of the undecylresorcarene (rccc isomer) with 5,6-dichloropyrazin-2,3-dicarboxylic acid imide gave (in analogy to similar examples reported before[60]) a cavitand bearing hydrogen bond donors and acceptor groups on its rim. In fact dimerization takes place leading to a large cylindrical capsule ($\approx$ 1.0×1.8 nm) stabilised by a "seam" of eight bifurcated hydrogen bonds as indicated in Fig. 9.26.[61] This dimerization was unambiguously proved by ^{1}H NMR for various examples, showing simple signals for the host, if suitable guests are present, and upfield shifted signals for the included guest(s).[62]

The cylindrical cavity favours the inclusion of elongated guest molecules, while mesitylene is obviously not included in a clean fashion, making mesitylene-d_{12} the solvent of choice for these studies (Fig. 9.27.). While a simple set of three singlets for the NH and ArH protons of the dimer is observed for symmetrical guests (e.g. dicyclohexyl carbodiimide or dibenzyl) less symmetrical guests such as benzyl phenyl ether or p-[N-(p-tolyl)]toluylamide **37** lead to a doubling of the host signals, since the two ends of the capsule are different. The guest obviously can spin along the main axis of the capsule, but is too large to tumble within it.[63]

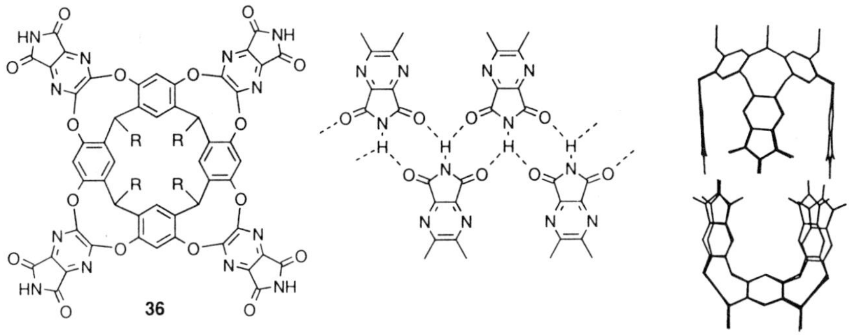

Figure 9.26. Structure of extended cavitand **36** (left), hydrogen bonding scheme (middle) and shape (right) of its dimer.

The capsules can distinguish between E and Z isomers of stilbene(s) with a selectivity of at least 50:1 (Fig. 9.28). For tertiary amides, such as **38**, which prefer the E-isomer, the E/Z equilibrium is shifted entirely towards the Z-isomer upon complexation/inclusion which is concluded from the analogy of the NMR-spectra with those obtained for the corresponding secondary amides (e.g. **37**) existing mainly as Z-isomer.[64]

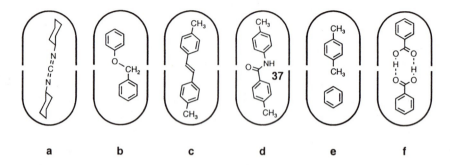

Figure 9.27. Typical inclusion complexes of dimers (**36**)₂. High field shifts are observed for all guests. For **b**, **d**, and **e** a double set of signals is observed for the host, showing the difference between both ends of the capsule.

Probably one of the most important properties of **37** is the simultaneous inclusion of two small guest molecules. In the case of 4-methylpyridine-2(1*H*)-one or benzoic acid (the larger 4-methylbenzoic acid is *not* included!) a hydrogen bonded dimer is encapsulated, leading to supermolecules of "second order". But also two toluene molecules or a benzene/*p*-xylene pair may be selectively included (compare Fig. 9.27.). In the latter case the two ends of the capsule become different again showing that (on the NMR-time scale) the two included molecules cannot pass each other. Encapsulation of two chiral guests from a racemic mixture leads to diastereomeric complexes. In the case of *trans*-1,2-cyclohexanediol the meso-complex, containing the two opposite enantiomers is favoured over the complex containing two molecules of the same enantiomer.

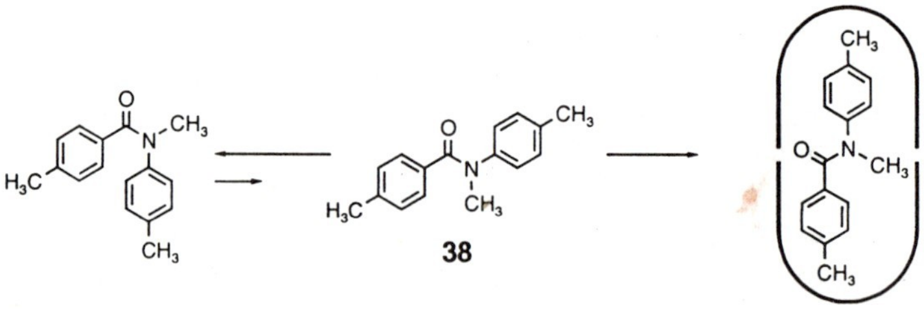

Figure 9.28. Shifting of the E/Z-equilibrium by encapsulation.

9.4. Outlook

The examples described above demonstrate the astonishing development which took place mainly during the past five years and which is taking place with permanently increasing speed while this book is edited. Still we are just on the beginning of an area the end of which cannot yet be foreseen, even not for the rather limited section of "calixarenes, self-assembling via hydrogen bonds".

Nevertheless, some developments can be predicted for the near future. Self-assembly will not be restricted to oligomeric species, but will enable the construction of well defined, three-dimensional polymeric architectures with controlled geometry (e.g. helicity). Encapsulation of molecules is just crossing the borderline from one to two and more guest molecules, and the use of such "capsules" as molecular reaction vessels is pre-designed. The introduction of additional functional groups will not only increase the selectivity of inclusion but will lead also to supramolecular catalysts. Learning more and more about the co-operativity of hydrogen bonding motifs, it will be possible also to design systems self-assembling in aqueous solutions, overcoming thus the "destructive" properties of this hydrogen bond breaking solvent. Like in natural systems guest inclusion will be favoured then by hydrophobic forces.

Some deliberate restriction was necessary for this chapter, but the entire potential of "self-assembled" artificial systems will be exhausted only by the combination of various structural elements or building blocks and by the intelligent use of various weak forces to control their ordered assembly.

Acknowledgement: We are very grateful to Profs. J. L. Atwood, J. Matthay, D. N. Reinhoudt, D. M. Rudkevich, and J. C. Sherman who made informations available to us before publication, helping thus, to make this review as actual as possible.

References

1. a) Lehn J.-M., *Supramolecular Chemistry. Concepts and Perspectives* (VCH, Weinheim, 1995). b) Philp D. and Stoddart J. F., *Angew. Chem., Int. Ed. Engl.* **35** (1996), 1154-1196.
2. Both macrocycles have the same $[1_n]$metacyclophane skeleton, see: Böhmer, V. *Angew. Chem., Int. Ed. Engl.* **34** (1995), 713-745.
3. a) Rebek J., Jr., *Chem. Soc. Rev.* **1996**, 255-264. b) Chapman R. G. and Sherman J. C., *Tetrahedron* **53** (1997), 15911-15945.
4. van Loon J.-D., Janssen R. G., Verboom W. and Reinhoudt D. N., *Tetrahedron Lett.* **33** (1992), 5125-5128.

5. a) Struck O., Verboom W., Smets W. J. J., Spek A. L. and Reinhoudt D. N., *J. Chem. Soc., Perkin. Trans.* 2 **1997**, 223-227; b) Arduini A., Fabbi M., Mantovani, M., Mirone L., Pochini A., Secchi A. and Ungaro, R. *J. Org.Chem.* **60** (1995), 1454-1457.
6. Arduini A., Domiano L., Ogliosi L., Pochini A., Secchi A. and Ungaro R., *J. Org. Chem.* **62** (1997), 7866-7868.
7. Koh K., Araki K. and Shinkai S., *Tetrahedron Lett.* **35** (1994), 8255-8258.
8. Vreekamp R. H., Verboom W. and Reinhoudt D. N., *J. Org. Chem.* **61** (1996), 4282-4288.
9. a) Murakami H. and Shinkai S., *J. Chem. Soc., Chem. Commun.* **1993**, 1533-1535; b) Murakami H. and Shinkai S., *Tetrahedron Lett.* **34** (1993), 4237-4240.
10. Lhotak P. and Shinkai S., *Tetrahedron Lett.* **36** (1995), 4829-4832.
11. a) Rudkevich D. M., Shivanyuk A. N., Brzozka Z., Verboom W. and Reinhoudt D. N., *Angew. Chem., Int. Ed. Engl.* **34** (1995), 2124-2126. b) See also: Shivanyuk A. N., Rudkevich D. M. and Reinhoudt D. N., *Tetrahedron Lett.* **37** (1996), 9341-9344.
12. Jakobi R. A., Böhmer V., Grüttner C., Kraft D. and Vogt W., *New J. Chem.* **20** 1996), 493-498.
13. Timmerman P., Verboom W., Reinhoudt D. N., Arduini A., Grandi S., Sicuri A. R., Pochini A. and Ungaro R., *Synthesis* **1994**, 185-189.
14. Shimizu K. D. and Rebek J., Jr., *Proc. Natl. Acad. Sci. USA* **92** (1995), 12403-12407.
15. Castellano R. K., Rudkevich D. M. and Rebek J., Jr., *J. Am. Chem. Soc.* **118** (1996), 10002-10003.
16. Mogck O., Böhmer V. and Vogt W., *Tetrahedron* **52** (1996), 8489-8496.
17. Mogck O., Paulus E. F., Böhmer V., Thondorf I. and Vogt W., *Chem. Commun.* **1996**, 2533-2534.
18. Frish L., Matthews S. E., Böhmer V. and Cohen Y., *J. Chem. Soc., Perkin Trans.* 2 **1999**, 669-671.
19. Schalley C. A., Castellano R. K., Brody M. S., Rudkevich D. M., Siuzdak G. and Rebek J., Jr., *J. Am. Chem. Soc.* **121** (1999), 4568-4579.
20. Castellano R. K., Kim B. H. and Rebek J., Jr., *J. Am. Chem. Soc.* **119** (1997), 12671-12672.
21. Hamann B. C., Shimizu K. D. and Rebek J., Jr., *Angew. Chem., Int. Ed. Engl.* **35** (1996), 1326-1329.
22. Mogck O., Pons M., Böhmer V. and Vogt W., *J. Am. Chem. Soc.* **119** (1997), 5706-5712.
23. O'Leary B. M., Grotzfeld R. M. and Rebek J., Jr, *J. Am. Chem. Soc.* **119** (1997), 11701-11702.
24. Castellano R. K. and Rebek J., Jr., *J. Am. Chem. Soc.* **120** (1998), 3657-3663.

25. Castellano R. K., Rudkevich D. M. and Rebek J., Jr., *Proc. Natl. Acad. Sci. USA* **94** (1997), 7132-7137.
26. Brody M. S., Schalley C. A., Rudkevich D. M. and Rebek J., Jr., *Angew. Chem., Int. Ed. Engl.* **38** (1999), 1136-1139.
27. a) Zerkowski J. A., Mathias J. P. and G. M. Whitesides G. M., *J. Am. Chem. Soc.* **116** (1994), 4305-4315. b) Zerkowski J. A. and Whitesides G. M., *J. Am. Chem. Soc.* **116** (1994), 4298-4304.
28. Whitesides G. M., Simanek E. E., Mathias J. P., Seto C. T., Chin D. N., Mammen M. and Gordon D. M., *Acc. Chem. Res.* **28** (1995), 37-44.
29. Timmerman P., Vreekamp R., Hulst R., Verboom W., Reinhoudt D. N., Rissanen K., Udachin K. A. and Ripmeester J., *Chem. Eur. J.* **3** (1997), 1823-1832.
30. Prins L. J., Timmerman P. and Reinhoudt D. N., *Pure & Appl. Chem.* **70** (1998), 1459-1468.
31. Jolliffe K. A., Crego Chalma M., Fokkens R., Nibering N. M. M., Timmerman P. and Reinhoudt D. N., *Angew. Chem., Int. Ed. Engl.* **37** (1998), 1247-1251.
32. Crego Chalma M., Fokkens R., Nibering N. M. M., Timmerman P. and Reinhoudt D. N., *Chem. Commun.* **1998**, 1021-1022.
33. Vreekamp R. H., van Duynhoven J. P. M., Hubert M., Verboom W. and Reinhoudt D. N., *Angew. Chem., Int. Ed. Engl.* **35** (1996), 1215- 1218.
34. Jolliffe K. A., Timmerman P. and Reinhoudt D. N., *Angew. Chem., Int. Ed. Engl.* **38** (1999), 933-937.
35. Klok H.-A., Jolliffe K. A., Schauer C., Spatz J. P., Möller M., Timmerman P., and Reinhoudt D. N., *J. Am. Chem. Soc.* **121** (1999), in press.
36. González J. J., Prados P. and de Mendoza J., *Angew. Chem., Int. Ed. Engl.* **38** (1999), 528-531.
37. Tunstad L. M., Tucker J. A., Dalcanale E., Weiser J., Bryant J. A., Sherman J. C., Helgeson R. C., Knobler C. B. and Cram D. J., *J. Org. Chem.* **54** (1988), 1305-1312.
38. a) MacGillivray L. R. and Atwood J. L., *J. Am. Chem. Soc.* **119** (1997), 6931-6932. b) The same structure was also published by: Iwanek W., Fröhlich, R., Urbaniak M., Näther C. and Mattay J., *Tetrahedron* **54** (1998), 14031-14040.
39. MacGillivray L. R. and Atwood J. L., *Chem. Commun.* **1999**, 181-182.
40. a) MacGillivray L. R., Holman K. T. and Atwood J. L., *Crystal Engineering* **1** (1998), 87-96. b) Ferguson G., Glidewell C., Lough A. J., McManus G. D. and Meehan P. R., *J. Mater. Chem.* **8** (1998), 2339- 2345.
41. a) Shivanyuk A., Paulus E. F., Böhmer V. and Vogt W., *Angew. Chem., Int. Ed. Engl.* **36** (1997), 1301-1303. b) Shivanyuk A, Paulus E. F., Böhmer V. and Vogt W., *Gazz. Chim. Ital.* **127** (1997), 741-747.

42. For a recent discussion of geometrical aspects of host molecules including self-assembled cavities see: MacGillivray L. and Atwood J. L., *Angew. Chem., Int. Ed. Engl.* **38** (1999), 1018-1033.
43. Rose K. N., Barbour L. J., Orr G. W. and Atwood J. L., *Chem. Commun.* **1998**, 407-408.
44. Murayama K. and Aoki K. *Chem. Commun.* **1998**, 607-608.
45. Kikuchu Y., Tanaka Y., Sutarto S., Kobayashi K., Toi H. and Aoyama Y., *J. Am. Chem. Soc.* **114** (1992), 10302-10306.
46. McGillivray L. R. and Atwood J. L., *Nature* **389** (1997), 469- 472.
47. Tanaka Y. and Aoyama Y., *Bull. Chem. Soc. Jpn.* **63** (1990), 3343-3344.
48. Gerkensmeier T., Iwanek W., Agena C., Fröhlich R., Kotila S., Näther C. and Mattay J., *Eur. J. Org. Chem.*, in press.
49. Sherman J. C., Knobler C. B. and Cram D. J., *J. Am. Chem. Soc.* **113** (1991), 2194-2204.
50. For a recent review on carcerands see: Jasat A. and Sherman J. C., *Chem. Rev.* **99** (1999), 931-967.
51. Chapman R. G. and Sherman J. C., *J. Am Chem. Soc.* **117** (1995), 9081-9082.
52. Chapman R. G., Olovsson G., Trotter J. and Sherman J. C., *J. Am. Chem. Soc.* **120** (1998), 6252-6260.
53. Chapman R. G. and Sherman J. C., *J. Org. Chem.* **63** (1998), 4103-4110.
54. Chapman R. G. and Sherman J. C., *J. Am. Chem. Soc.* **120** (1998), 9818-9826.
55. Sherman J. C., personal communication.
56. Chopra N. and Sherman J.C., *Angew. Chem., Int. Ed. Engl.* **36** (1997), 1727-1729.
57. Chapman R. G. and Sherman J. C., *J. Am. Chem. Soc.* **121** (1999), 1962-1963.
58. Ma S., Rudkevich D. M. and Rebek J., Jr., *J. Am. Chem. Soc.* **120** (1998), 4977-4981.
59. Mecozzi S. and Rebek J., Jr., *Chem. Eur. J.* **4** (1998), 1016-1022.
60. Moran J. R., Ericson J. L., Dalcanale E., Bryant J. A., Knobler C. B. and Cram D. J., *J. Am. Chem. Soc.* **113** (1991), 5707-5715.
61. Heinz T., Rudkevich D. M. and Rebek J., Jr., *Nature* **394** (1998), 764-766 .
62. For intramolecular hydrogen bonding in a similar cavitand, controlling the guest exchange see: Rudkevich D. M., Hilmersson G. and Rebek J., Jr., *J. Am. Chem. Soc.* **119** (1997), 9911-9912.
63. Compare the „carcerism" described by Timmerman P., Verboom W., van Veggel F. C. J. M., van Duynhoven J. P. M. and Reinhoudt D. N., *Angew. Chem., Int. Ed. Engl.* **33** (1994), 2345-2348.
64. Heinz T., Rudkevich D. M. and Rebek J., Jr., *Angew. Chem., Int. Ed. Engl.* **38** (1999), 1136-1139.

CHAPTER 10

CALIXARENE BASED CATALYTIC SYSTEMS

ROBERTA CACCIAPAGLIA and LUIGI MANDOLINI

Centro CNR Meccanismi di Reazione and Dipartimento di Chimica Università La Sapienza, Box 34 Roma 62, 00185 Roma, Italy.

10.1. Introduction

In recent years there has been a growing interest in the rational design of synthetic catalysts capable of achieving significant results in terms of reaction rate and selectivity. No doubt, the admirable properties of enzymes have provided chemists with a formidable challenge and a standard to be emulated. A distinctive feature of enzyme catalysis is substrate (S) binding to form a Michaelis-Menten complex ($S{\cdot}cat$), followed by transformation of the bound substrate by means of strategically placed catalytic groups (Scheme 10.1). The several features possessed by the calixarenes, notably their ability to form host-guest complexes and the possibility of introducing a large variety of functions by means of selective derivatization, qualify this class of macrocycles as good candidates to act as enzyme mimics or, more in general, as synthetic catalysts.

This chapter aims at describing the various ways calixarenes have provided the basis for rational catalyst design. Following a short section on the use of calixarene derivatives in phase transfer catalysis, a number of examples will be discussed in

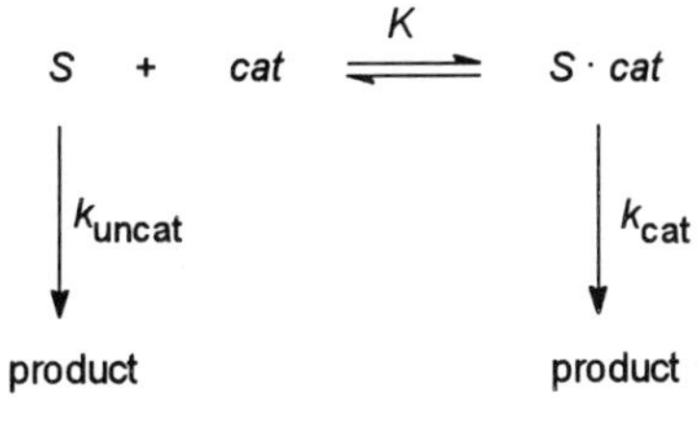

$$S + cat \underset{}{\overset{K}{\rightleftharpoons}} S \cdot cat$$

$$\downarrow k_{uncat} \qquad\qquad \downarrow k_{cat}$$

$$\text{product} \qquad\qquad \text{product}$$

Scheme 10.1

which substrate binding to the calixarene cavity is believed to be crucial to the catalysis. The remaining part of the chapter is devoted to metal complexes of calixarenes. For convenience, metal complexes using the four calix[4]arene oxygen atoms as ligands are discussed separately. The last section illustrates the use of either rim of the calixarene framework as a convenient platform for the introduction of metal ion binding sites and catalytic groups.

10.2. Phase Transfer Catalysis

Most of the work in this area has been done by Japanase workers. The octopus-type calix[6]arene **1**, endowed with cation binding properties thanks to the six 3,6,9-trioxadecyl chains, was first developed by Funada et al.[1] and utilised by Nomura and his group as an efficient phase transfer catalyst in several reaction systems including (*i*) the formation of ethers from phenols and alkyl or aryl halides in the presence of solid KOH (Eq. 10.1);[2] (*ii*) the ester forming reaction of alkali metal carboxylates with alkyl halides (Eq. 10.2);[3] (*iii*) the generation of dichlorocarbene from $CHCl_3$/KOH (Eq. 10.3) and its subsequent addition to alkenes and alkadienes;[4] and (*iv*) the oxidation of alkenes, alkynes, and alcohols with $KMnO_4$ (Eq. 10.4).[5] The common features shared by these reaction systems is that an alkali metal salt of an anionic reactant is involved. Convincing evidence has been produced that the catalytic ability of **1** is determined primarily by the binding ability toward the metal cation and, consequently, by the anion solubilizing and

$$\text{ArOH} + \text{R-X} \xrightarrow[\text{CH}_2\text{Cl}_2]{cat\,/\,\text{KOH}} \text{ArOR} \qquad (10.1)$$

$$\text{RCOOM} + \text{R'-X} \xrightarrow[\text{CH}_2\text{Cl}_2]{cat} \text{RCOOR'} \qquad (10.2)$$

$$\text{CHCl}_3 \xrightarrow[\text{CH}_2\text{Cl}_2]{cat\,/\,\text{KOH}} :\text{CCl}_2 \xrightarrow{\text{alkene}} \text{Adduct} \qquad (10.3)$$

$$\text{RCH=CH}_2 \text{ (or RC≡CH or RCH}_2\text{OH)} \xrightarrow[\text{CH}_2\text{Cl}_2]{cat\,/\,\text{KMnO}_4} \text{RCO}_2\text{H} \qquad (10.4)$$

activating properties derived therefrom. Most reactions have been successfully carried out in CH_2Cl_2, either with or without addition of small amounts of water. In many instances the effectiveness of the calixarene catalyst is comparable or even greater than that of 18-crown-6 or tetraalkylammonium salts. Interestingly, the dichlorocarbene generated in the presence of **1** displays selectivity features different from those observed when 18-crown-6 is used as catalyst.[4] Thus, from *d*-limonene (Eq. 10.5) the mono-adduct was the major product, along with small amounts of the bis-adduct, when the dichlorocarbene was generated in the presence of the calixarene catalyst. But in the presence of 18-crown-6, the bis-adduct was the major product. Similar results were obtained with other alkadienes. It has been suggested that the calixarene-catalysed addition of dichlorocarbene to the alkadiene occurs within the cavity of the calixarene.

The design of the ionophoric calix[4]arene analogue **2**,[6] bearing four 3,6,9-trioxadecyl units, was clearly inspired by **1**. Its phase transfer catalytic properties were compared to those of **1** and its tetrameric analogue **3** using the reaction between phenol and benzyl bromide in either CCl_4 or CD_2Cl_2 in the presence of alkali metal hydroxides. It was found that the catalytic effectiveness of **2** was comparable to that of **1** and slightly higher than that of **3**. A study by Shinkai et al.[7] provides a critical evaluation of the structural factors affecting the phase transfer catalytic capability of calix[4]arenes. The authors compared the activity of the *cone* isomers of calix[4]arenes **4-6** in the ester forming reaction of $C_3H_7CO_2Na$ with $p\text{-}O_2NC_6H_4CH_2Br$ in a water/CH_2Cl_2 two-phase system. The conclusion was reached that lipophilic groups bulkier than *tert*-butyl introduced into the upper rim enhance the catalytic activity and that there is no need for a 3,6,9-trioxadecyl chain which, in fact, causes emulsification during work-up. The four OCH_2CH_2OMe chains of **6** suffice to compose an effective ionophoric cavity.

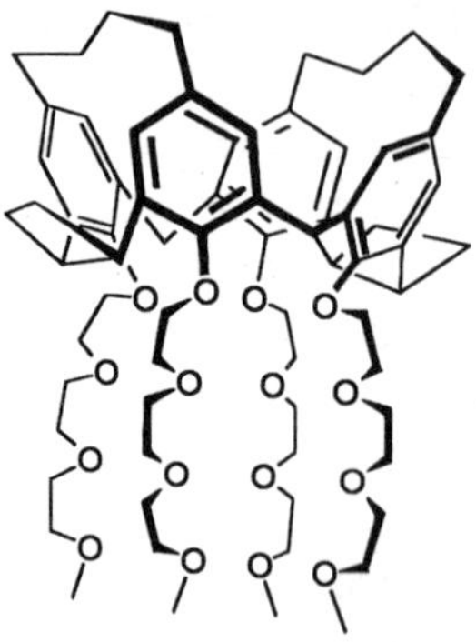

Instead of ethereal oxygens, the carbonyl oxygens of amide functions can be used as cation complexing agents. The *p-tert*-butylcalix[4]arene tetramide **7**[8] has been found to be an effective phase transfer catalyst in the dehydrohalogenation of 3-halopropionamides to afford *N*-substituted azetidin-2-ones (Eq. 10.6). Improved yields of azetidin-2-one products were obtained in several instances when compared with the 18-crown-6 catalyst.

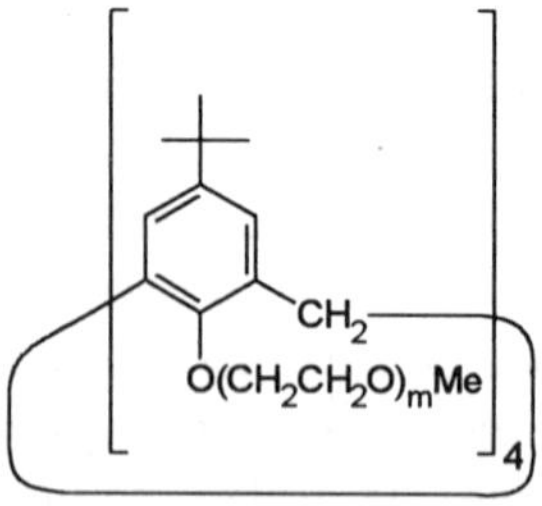

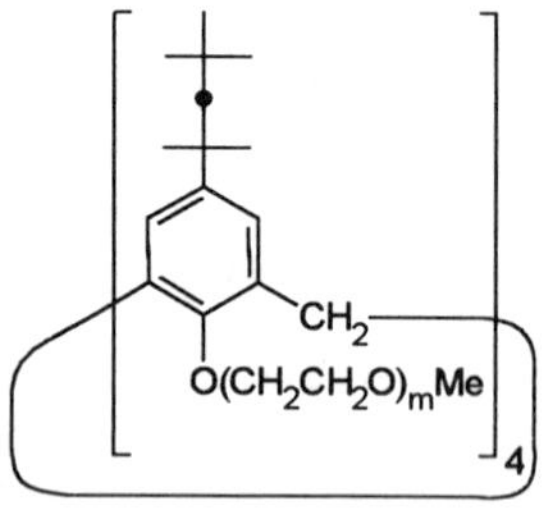

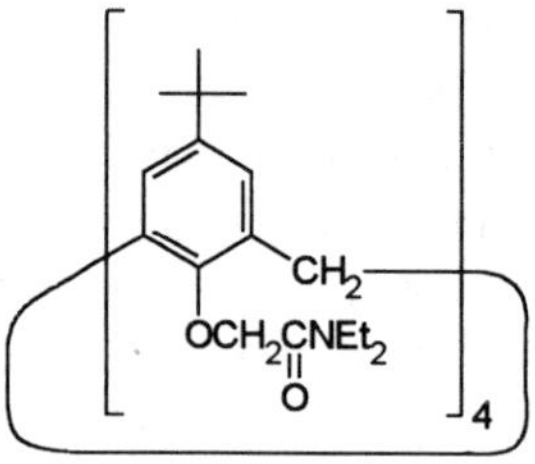

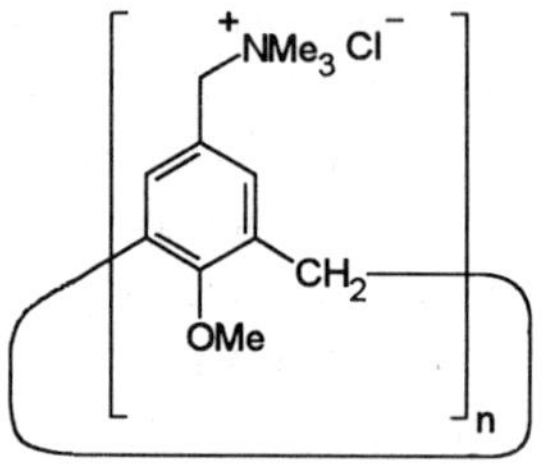

The water-soluble calixarenes **8-10**[9] containing trimethylammoniomethyl groups act as efficient inverse phase transfer catalysts in the reaction of sodium cyanide in water solution with a series of alkyl halides, including 1-bromooctane, 1-iodooctane, β-phenethyl bromide, benzyl bromide and chloride, and

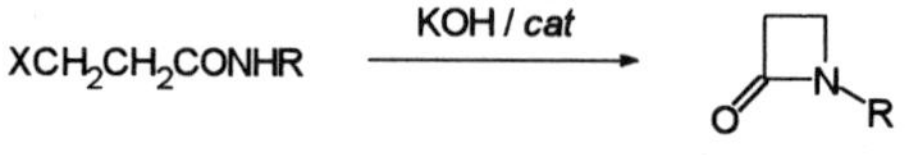

$$XCH_2CH_2CONHR \xrightarrow{KOH\ /\ cat} \underset{O}{\overset{}{\square}}\!\!-\!N\!-\!R \qquad (10.6)$$

2-(bromomethyl)naphthalene. No significant difference in activity was detected in the reaction of benzyl bromide, but with the bulkier 2-(bromomethyl)naphthalene the order is **10>9>8**, thus suggesting that some sort of recognition process ruled by the hole-size relationship plays an important role.

10.3. Catalytic Receptors

To act in water solution, a host molecule must possess both a solubilizing hydrophilic site and a hydrophobic cavity to bind a substrate. The hexasulfonated calix[6]arenes synthesized by Shinkai et al[10] satisfy the above requisites in that they are water-soluble and bind small molecules in water. Derivatives **11** (n=6) and **12**, endowed with acidic groups at the lower rim, efficiently catalyse the hydration of 1-benzyl-1,4-dihydronicotinamide (BNAH), a synthetic analogue of NADH, for which rate-determining protonation has been suggested (Eq. 10.7).[11] There is little doubt that the catalysis is a real *endo*-calix process, as shown by the good adherence of the kinetics to Scheme 10.1 (Michaelis-Menten kinetics), with binding constants derived from analysis of rate data (30 °C, pH 6.3) consistent with spectroscopic fluorescence data. Shinkai believes that the catalysis is the result of a bifunctional action wherein a proton is transferred from an acidic group at the lower rim and the developing positive charge is stabilised by electrostatic interaction with the anionic groups at the upper rim (Fig. 10.1), in agreement with the suggested mechanism for the hydration of NADH catalysed by

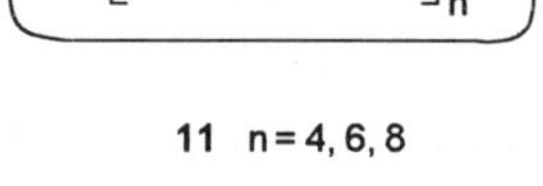

11 n = 4, 6, 8 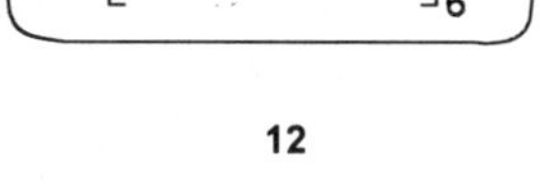   **12**

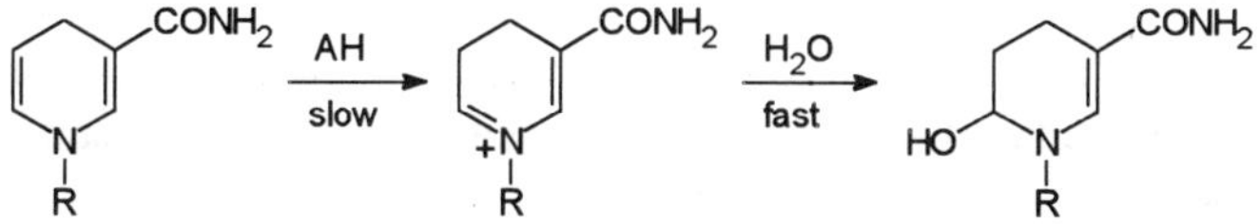

(10.7)

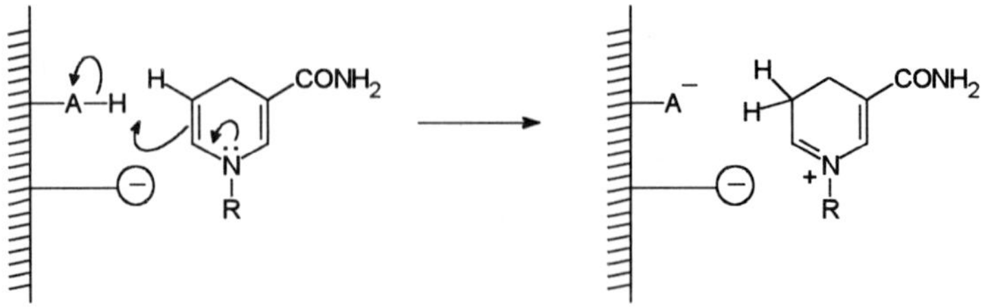

Figure 10.1. Hydration of NADH and its synthetic analogues. Mechanism of the rate-limiting protonation step.

glyceraldehyde-3-phosphate dehydrogenase.[11] An analogous study by Gutsche and Alam[12] of the hydration of BNAH catalysed by p-(carboxyethyl)calix[n]arenes **13** (n=4-8) has shown that the calix[6]arene is a better catalyst than either its larger or smaller analogues, but substantially less effective than the sulfonated calix[6]arene **11** (n=6).

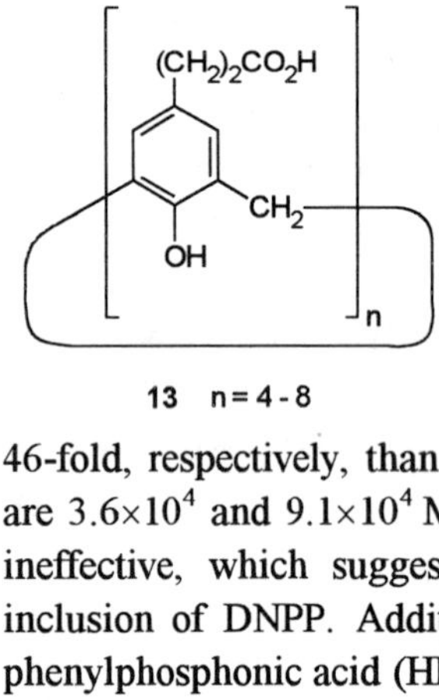

The cationic groups located at the upper rim of the water soluble calix[6]arenes **14** provide the driving force for binding to the 2,4-dinitrophenyl phosphate dianion (DNPP) and its subsequent hydrolysis. In a kinetic study carried out at pH 10, 30 °C, Shinkai et al[13] observed saturation kinetics, with catalytic rate constants for R=Me and C_8H_{17} greater by 21-fold and 46-fold, respectively, than in bulk water. The corresponding binding constants K are 3.6×10^4 and 9.1×10^4 M^{-1}. The tetrameric analogue **15** was found to be almost ineffective, which suggests that its molecular architecture is not suitable for inclusion of DNPP. Addition of the competitive inhibitors phosphoric acid and phenylphosphonic acid (HPO_4^{2-} and $PhPO_3^{2-}$, respectively, at pH 10) suppressed the catalytic activity of the calix[6]arene catalysts. Another reaction which is catalysed by the calix[6]arenes **14**, R = Me, C_8H_{17} is the basic hydrolysis of p-nitrophenyl dodecanoate.[14] The catalytic rate constants, measured at pH 8.51, 30 °C, are more than five orders of magnitude larger than in the presence of **15**. The dodecanoate ester substrate is known to form aggregates in water in which the ester group is protected from nucleophilic attack by OH$^-$. The "host-guest" mechanism of catalysis for the reaction catalysed by **14** is argued to be much more effective than the "deshielding" mechanism, caused by simple quaternary salts such as N,N,N-trimethylanilinium iodide, as well as by **15** whose basket is not large enough for the host-guest mechanism to be operative.

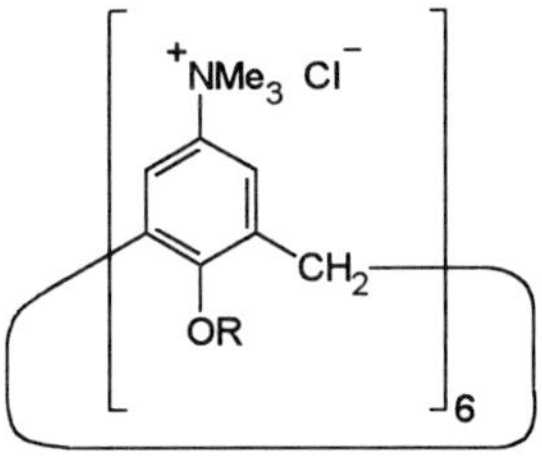

14 R = Me, C$_8$H$_{17}$ 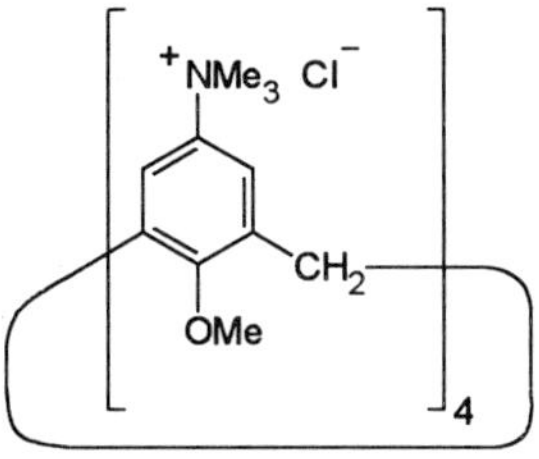   **15** 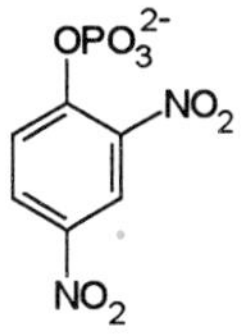   DNPP

Calix[n]arenes **11**, n=4, 6, and 8 have been reported[15] to act as artificial ribonucleases in the ring opening hydrolysis of cytidine-2',3'-cyclic phosphate at pH 2, 25 °C. Modest rate enhancements and regioselectivity are induced by these calixarenes, the largest values being a 6-fold rate acceleration and a 3'/2' regioselectivity increase from 1.5 to 2.7 with the calix[4]arene catalyst. Poorer results were obtained in the hydrolysis of 2',3'-cyclic phosphates of adenosine, guanosine and uridine. The catalysis involves complex formation between cytidine-2',3'-cyclic phosphate and the calix[4]arene, as indicated by saturation kinetics, and confirmed by ^{1}H NMR spectroscopy. Electrostatic interactions and hydrogen bonding have been suggested to provide the driving force for the formation of the complex, whose proposed structure is shown in Fig. 10.2. Thus, the catalytic mechanism involves an *exo*-calix process, unlike the preceding examples in this section.

An elegant example of complexation catalysis based on a resorcinarene derivative, the only one involving a member of this class of macrocycles, has been reported by Williams and coworkers.[16] Their artificial esterase **16** combines a concave binding site with a number of pendant arms bearing dimethylamino

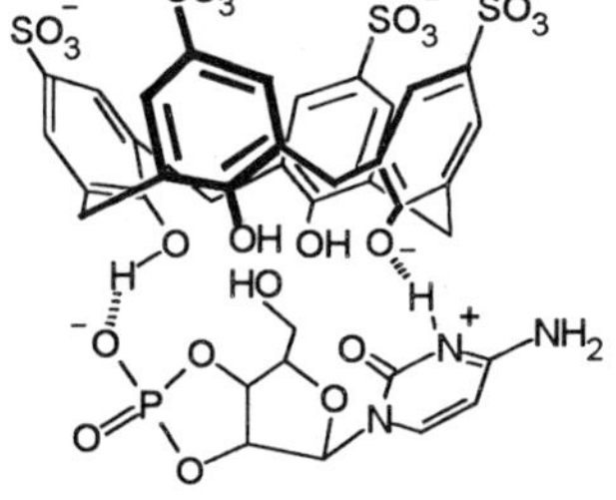

Figure 10.2. Proposed structure of the complex of **11**, n=4 with cytidine-2',3'-cyclic phosphate at pH 2.

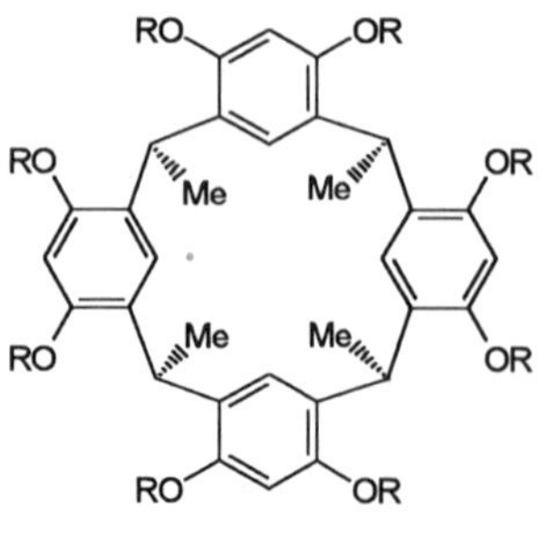

16, R = CH$_2$CONH(CH$_2$)$_3$NMe$_2$

catalytic functions. Catalysis of the cleavage of p-nitrophenyl esters of a large variety of carboxylic acids in moderately alkaline water solution strictly adheres to Michaelis-Menten kinetics (Scheme 10.1). The suggested mechanism involves complexation of the substrate with host **16**, followed by rate-limiting intracomplex nucleophilic attack to form a reactive N-acylammonium intermediate which decomposes rapidly to form acid and regenerate the amine catalyst. Comparison of the catalytic rate constant k_{cat} (0.011 s^{-1}, pH>9.6) for p-nitrophenyl acetate with the rate constants for the analogous intramolecular reactions of the p-nitrophenyl esters of 4-N,N-dimethylaminobutyric (358 s^{-1}) and 5-N,N-dimethylaminovaleric (1.67 s^{-1}) acids, reveals but a modest efficiency of the intracomplex reaction. This is understandable because the substrate complexed with the host does not have a constricted conformation in which the ester function is held in close proximity of the amino group.

Addition of either a calixarene or resorcinarene host to an ester substrate undergoing hydrolysis not always results in rate acceleration. Chawla and Pathak[17] have found that the basic hydrolysis of phenyl benzoates is inhibited by p-alkyl-calix[n]arenes with n=4-8, the strongest inhibition being observed with p-*tert*-butylcalix[8]arene. Strong retardation for the basic hydrolysis of

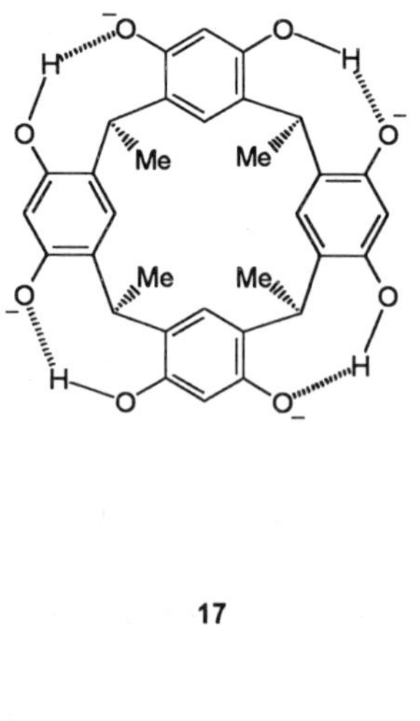

17 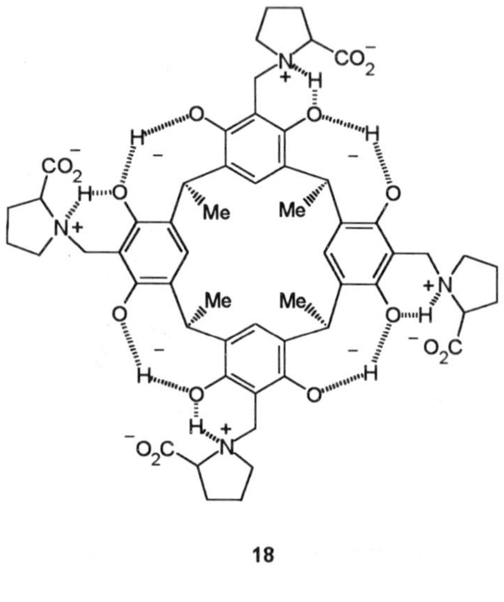   **18**

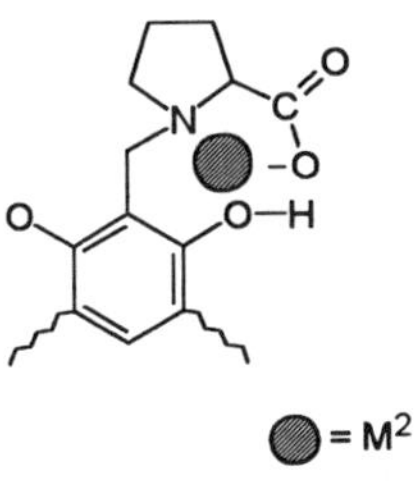

acetylcholine (60 °C, pH 10) is caused by complexation with the bowl-shaped tetraanion **17**.[18] Rate inhibition is believed to arise from a diminution of the activating effect of the positive charge of the choline moiety by the four-fold negative charge of the receptor. Such an inhibition completely disappears with the tetra L-proline derivative **18**.[19] The tetra-Zn^{II} complex **19**, was synthesized by Schneider and Schneider[19] with the idea in mind that the complexed metal ions would catalyse the hydrolysis of complexed acetylcholine via a Lewis acid mechanism or water activation. However it was found that the hydrolysis of acetylcholine at pH 10, 60 °C, in the presence of the Zn^{II} complex **19** takes place at the same rate as in the presence of the ligand **18** alone. Similar results were obtained with the Cu^{II} complex.

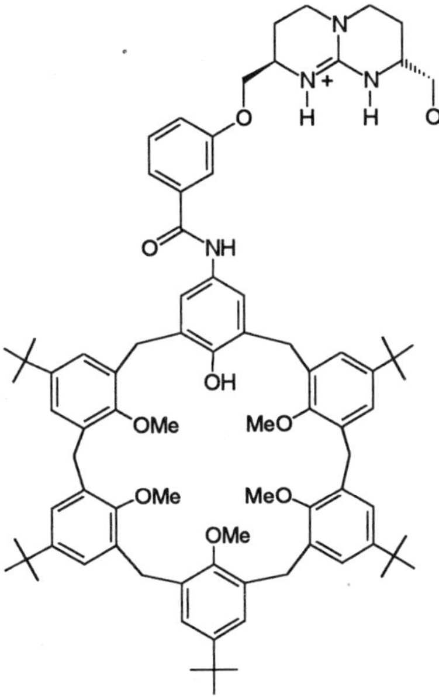

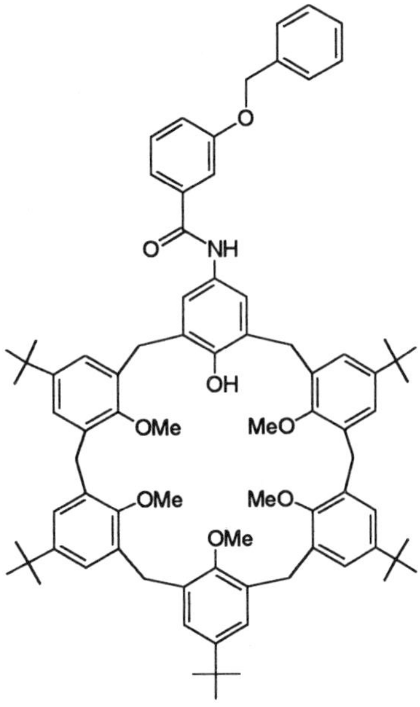

A more successful attempt at a synthetic acetylcholinesterase analogue is based on the ditopic receptor **20** synthesized by de Mendoza et al.,[20] that mimics the phosphocholine binding site of the McPC603 antibody. Receptor **20** binds strongly to DOPC, a transition state analogue for the hydrolysis of esters and carbonates, as a result of combined binding of the phosphate monoanion to the guanidinium moiety and of the choline trimethylammonium group to the calix[6]arene subunit. The methanolysis of choline p-nitrophenyl carbonate (p-$O_2NC_6H_4OCOOCH_2CH_2NMe_3I$) in $CHCl_3$-MeOH 99:1 (v/v) in the presence of diisopropylethylamine-perchlorate salt buffer is strongly catalysed by receptor **20**, but much less effectively by the monotopic model compounds **21** and **22** (Table 10.1).[21] Independent ^1H NMR experiments indicate that 13% of the choline carbonate is complexed to **20** under the conditions of the kinetic experiments reported in Table 10.1. Receptor **20** is a genuine turnover catalyst, because in its presence a 4- to 20-fold excess of substrate undergoes complete methanolysis at a rate that is significantly higher than background. The catalytic process is strongly inhibited by the addition of DOPC.

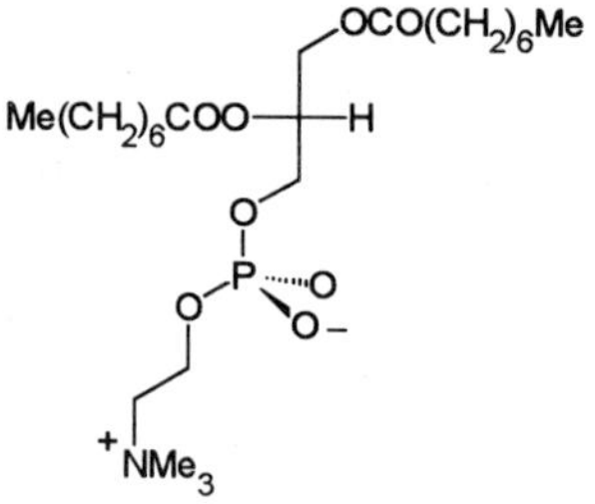

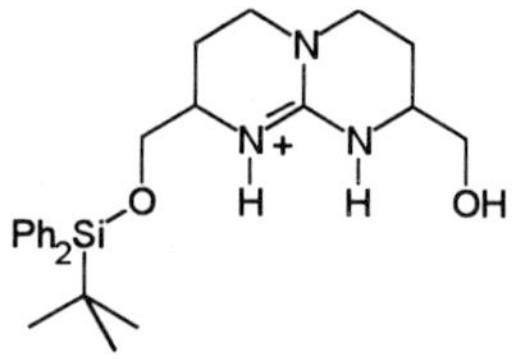

Table 10.1. Effect of 1.0 mM additives on the rate of methanolysis of 50 μM choline p-nitrophenyl carbonate in $CHCl_3$-MeOH 99:1 v/v in the presence of 25 mM diisopropylethylamine 0.50 mM perchlorate salt buffer at 25 °C.

Additive(s)	k_{obs} (s^{-1})	k_{rel}
none	1.85×10^{-4}	1.0
21	3.80×10^{-4}	2.1
22	1.67×10^{-3}	9.0
21+22	1.64×10^{-3}	9.0
20	1.41×10^{-2}	76

10.4. Calix[4]arenes as Tetradentate Ligands

The use of calix[n]arenes as polyoxo matrixes to support reactive metals capable of driving catalytic or stoichiometric reactions is a very young and largely unexplored area, restricted so far to metal complexes of *p-tert*-butylcalix[4]arene **23**, and of its mono- and dimethyl derivatives **24** and **25**, respectively.

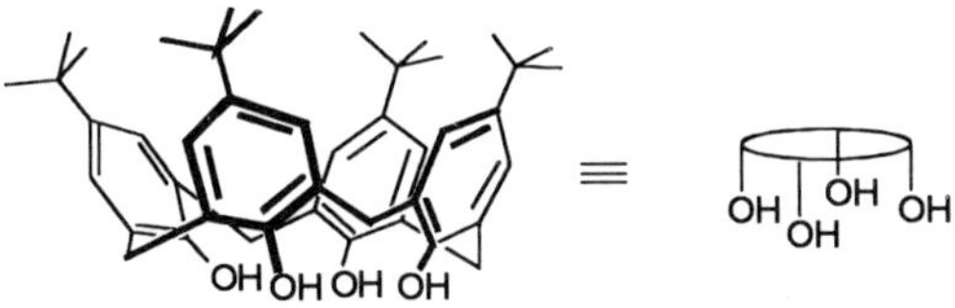

A genuine example of catalyst of this kind is provided by the substituted chelate phenolatoaluminum chloride **26**, derived from the reaction of **25** with diethylaluminium chloride (Eq. 10.8).[22] Metal complex **26** was successfully applied

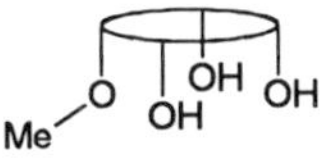

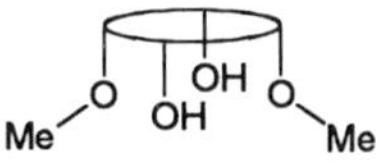

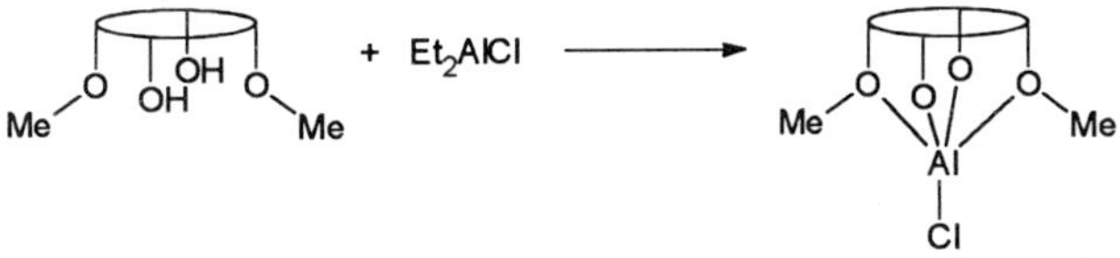

$$\text{25} + Et_2AlCl \longrightarrow \text{26} \tag{10.8}$$

to the polymerisation of propylene oxide and cyclohexene oxide in toluene under homogeneous conditions. Whereas poly(propylene oxide) obtained in the presence of the above catalyst contained isotactic diads predominantly, poly(cyclohexene oxide) did not show prevailing sequences of either isotactic or syndiotactic diads.

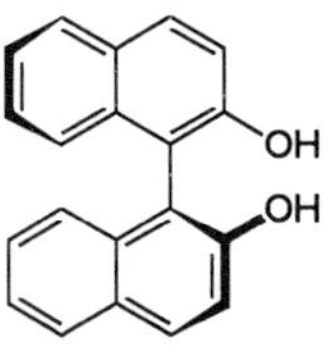

The activity of the zirconium-BINOL catalyst developed by Umani-Ronchi and coworkers[23] for the enantioselective allylation of aldehydes with allyltributyltin (Eq. 10.9) is significantly enhanced by calix[4]arene **23**. Addition of **23** in quantities as small as 0.5% increased the *ee* from 53 to 95% with linear aldehydes. Good results were obtained also with

$$\text{RCHO} + \diagup\!\!\!\diagdown\!\!\text{SnBu}_3 \xrightarrow[\text{p-tert-butylcalix[n]arene}]{(S)\text{-BINOL-ZrCl}_4(\text{thf})_2} \underset{R}{\overset{OH}{\diagup\!\!\!\diagdown\!\!\diagdown}} \qquad (10.9)$$

aromatic aldehydes. Calix[6]arene and calix[8]arene showed activities comparable to that of the cyclic tetramer, but other simple mono- and diphenols were much less effective. It is unfortunate that no information has been reported on the structure of the species responsible for this remarkable chiral-achiral ligand synergy.

Extensive studies of the use of calix[4]arene 23 and its derivatives 24 and 25 as polyoxo matrixes for reactive organometallic functionalities have been carried out in recent years by the Floriani group.[24] The role of the calix[4]arene ligand is unique in that the properties imparted to the anchored transition metal are different from those found in the corresponding complexes of the same number of monomeric phenoxo units. This is due to the fact that the quasiplanar geometry of the four oxygen donors has a major effect in determining the set and the relative energy of the frontier orbitals available at the metal centre.[24] An additional advantage is that the total charge of the O_4 moiety can be easily adjusted from -4 to -2 via partial methylation, thus providing the possibility of controlling the functionalisation degree of the metal. Metal-driven reactions in the unique environment provided by the calix[4]arene ligand are useful models of metal reactivity on oxo surfaces. Since the matter has been recently reviewed by Floriani himself,[24] only a few examples of the remarkable chemistry taking place at the metal centre of calix[4]arene-metal complexes 27-30 will be reported here.

A variety of C-C bond formation processes are assisted by zirconium bonded to the dimethoxycalix[4]arene dianion derived from 25. The metal-carbon bonds of the $Zr(Me)_2$ fragment of 27 undergo multistep migratory insertion of carbon monoxide and isocyanides (Eqs. 10.10 and 10.11).[25,26]

(10.10)

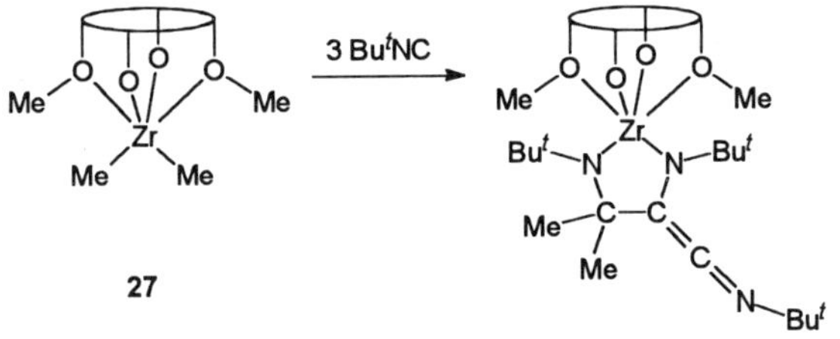

(10.11)

Migratory insertion reactions of carbon monoxide and isocyanides have also been observed with the calix[4]arene tantalum (V) derivative **28** (Eqs. 10.12 and 10.13).[27] Additional migratory insertion reactions involving the tantalum-butadiene fragment in **29** are shown in Eqs. 10.14 and 10.15.

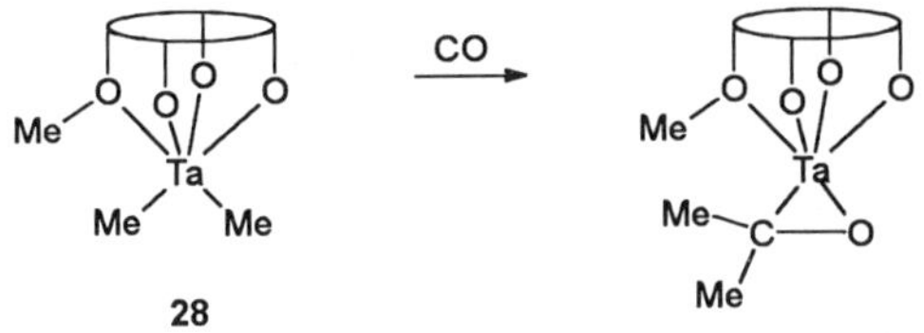

(10.12)

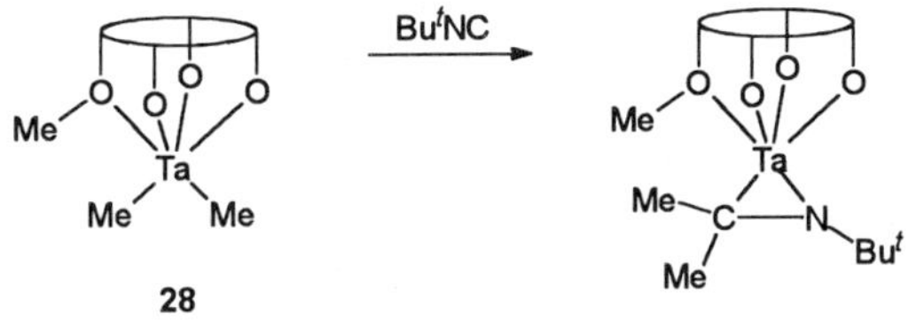

(10.13)

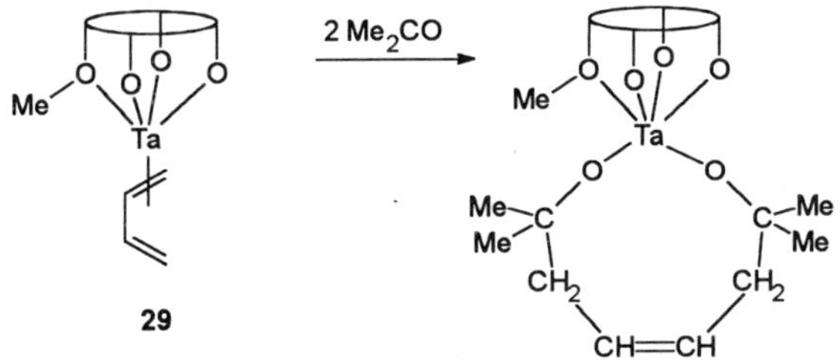

(10.14)

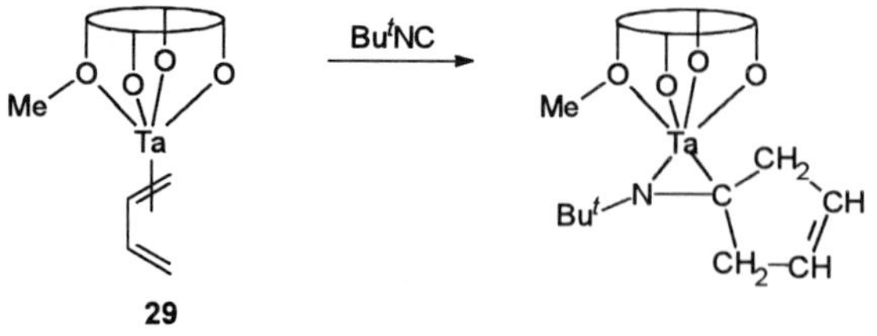

$$\text{(Scheme 10.15)}$$

Complexation of olefins by the tungsten (IV)-calixarene fragment has been achieved by the reaction of **30** in the presence of the appropriate olefin (Eq. 10.16).[28] The complexed ethylene moiety undergoes deprotonation with BuLi, leading to the corresponding anionic alkylidyne. The latter can be protonated with pyridinium chloride to the corresponding alkylidene. The overall result of the deprotonation-protonation sequence is the isomerisation of ethylene to ethylidene. Tungsten alkylidyne derivatives can also be prepared by treatment of **30** with 3 mol equiv of Li or Mg alkylating agents.[29]

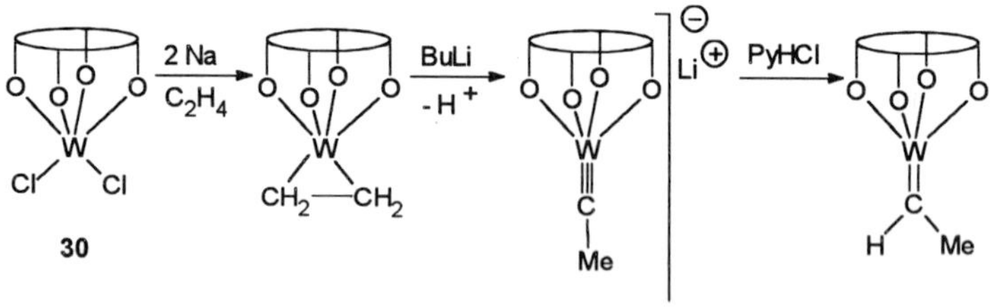

$$\text{(Scheme 10.16)}$$

Although the above processes are stoichiometric rather than catalytic, Floriani believes[24] that the preorganised oxo arrangement of calix[4]arene anions has the potential for metal complexes which can challenge the well known heterogeneous oxo-catalysts in a number of synthetically useful reactions.

10.5. Calixarenes as Catalytic Platforms

There is a considerable mass of evidence[30] that alkaline-earth metal ions are efficient promoters of acyl transfer reactions from esters to anionic nucleophiles, and that the observed rate enhancements are particularly large when the ester substrate is covalently linked to a metal ion binding site. This notion is well illustrated by the report[31] that the half-life for the basic methanolysis of the monoacetate of *p-tert*-butylcalix[4]arene-crown-5 **31** at 25 °C is reduced from

255

[Structures 31 and 32 shown]

31   32

34 weeks to 8 seconds by the addition of Ba^{2+} ion. Clearly, the lower rim of calix[4]arene offers a unique opportunity for the construction of a reaction site in which the metal ion, held in place by the crown-ether moiety and by the ionised hydroxyl supplies electrophilic activation to the ester carbonyl toward nucleophilic attack, as schematically shown in 32. This phenomenon was conveniently put to the purpose of developing a nucleophilic catalyst with transacylase activity.[32,33] An equimolar mixture of 31 and barium salt was found to act as a turnover catalyst in the methanolysis of *p*-nitrophenyl acetate and other aryl acetates in MeCN-MeOH (9:1, v/v, 25 °C) in the presence of diisopropylethylamine buffer. As shown in Fig. 10.3, the slow liberation of *p*-nitrophenol due to background methanolysis is unaffected by the addition of 31, but is strongly accelerated by a 1:1 mixture of 31 and BaBr$_2$. The initial burst of *p*-nitrophenol release is followed by a slow linear phase. These biphasic kinetics are typical of the proteolytic enzymes (chymotripsin) and are consistent with a double displacement (ping-pong) mechanism in which the catalyst takes the acetyl from the ester and transfers it to methoxide ion, thus restoring its active form and turning over (Eq. 10.17). Accumulation of the acetylated intermediate was monitored by HPLC analysis, and found to be fully consistent with the kinetics. The Sr^{2+} complex of 31 was found to be slightly less

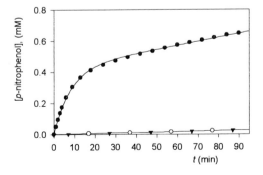

Figure 10.3. Liberation of *p*-nitrophenol in the methanolysis of *p*-nitrophenyl acetate. (▼) Background reaction in the presence of buffer alone; (○) buffer plus 0.46 mM 31; (●) buffer plus 0.46 mM 31 and 0.46 mM BaBr$_2$.

$$\text{ArOAc} + [\text{Ba}^{2+} \text{complex}] \longrightarrow \text{ArOH} + [\text{Ba}^{2+}\text{-OAc complex}] \xrightarrow{\text{MeO}^-} \text{AcOMe} + [\text{Ba}^{2+} \text{complex}] \qquad (10.17)$$

efficient than the Ba^{2+} complex, whereas the Ca^{2+} complex could not be investigated because of solubility problems.[34] Any catalytic activity disappears when the polyether bridge spans two adjacent oxygens of **23**, or when it is replaced by two distal methoxy groups.[34] The crucial importance of a certain degree of flexibility in the calix[4]arene platform is shown by the lack of any catalytic activity of the barium complexes of **33** and **34**.[35] In the methanolysis of a series of aryl acetates catalysed by the barium complex of **31**, a gradual changeover from rate-determining acylation (with the most reactive *p*-nitrophenyl acetate) to rate-determining deacylation (with the least reactive phenyl acetate) was observed.[33] It was shown that the scope of the catalyst is practically restricted to esters less reactive than *p*-nitrophenyl acetate and more reactive than phenyl acetate. It was also shown that breakdown of the tetrahedral intermediate is the slow step in the acylation of the catalyst. Preliminary evidence has been obtained[33] that the neighbouring protonated amino group in **35** can assist as a general acid the departure of the aryloxide leaving group from the tetrahedral intermediate.

Several 1,2- and 1,3-crown ether derivatives of *p-tert*-butylcalix[5]arene have been studied as potential catalysts of acyl transfer processes.[36] The methanolysis of *p*-nitrophenyl acetate was catalysed by the Ba^{2+} complexes of 1,3-crown ether derivatives only. The crown-5 compound **36** was found to be more effective than its crown-6 and crown-7 homologues, but substantially less effective than the corresponding calix[4]arene catalyst **31**.

33   **34**

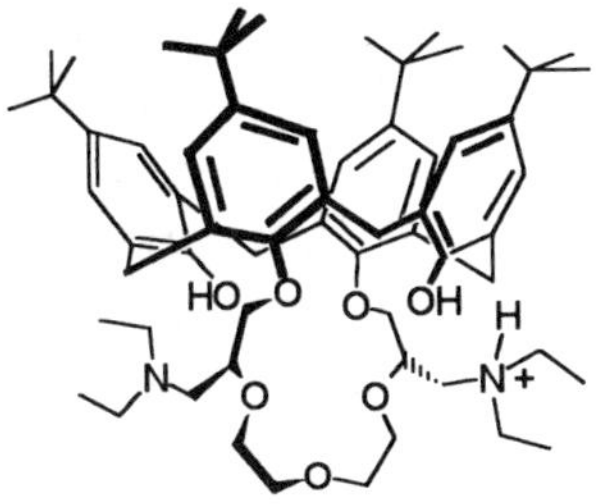

35 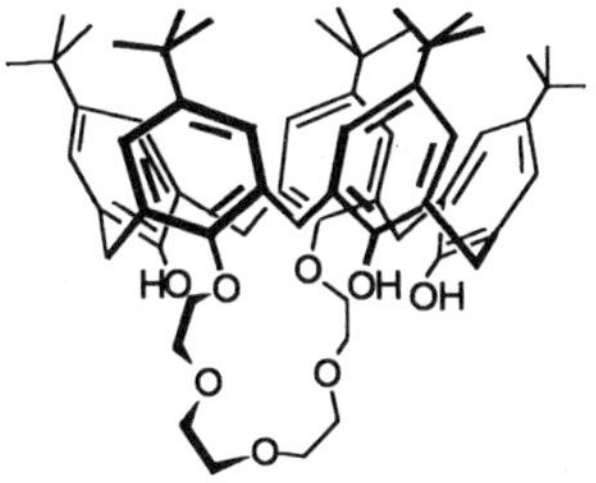 **36**

The Rh^I complex **37** of the diamide-diphosphane compound obtained via exhaustive functionalisation of the lower rim of **23** has been synthesized by Loeber et al.[37] Complex **37** catalyses the hydroformylation of styrene to 2-phenylpropanal and 3-phenylpropanal in a 95:5 ratio, which is not unusual in rhodium-phosphane catalysis. The low reaction rate observed was ascribed to the difficulty of approach of the substrate to the metal due to the steric bulk of the ligands and of the amide pendant groups.

The efficient Rh^I catalysts very recently reported by Paciello et al.,[38] provide an elegant example of catalyst design by lower rim modification of calix[4]arene. Ligand **38** is the first compound of a new class of chelate phosphites in which two phosphorous atoms are directly bonded to the two adjacent oxygens of the calix[4]arene platform. The hydroformylation of 1-octene with [Rh(CO)$_2$(acac)]/**38** proceeds slowly but with a selectivity of 200:1 in favour of the linear nonanal

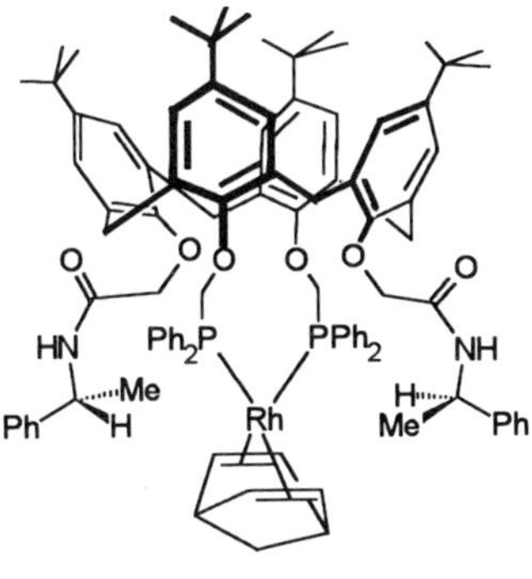

37 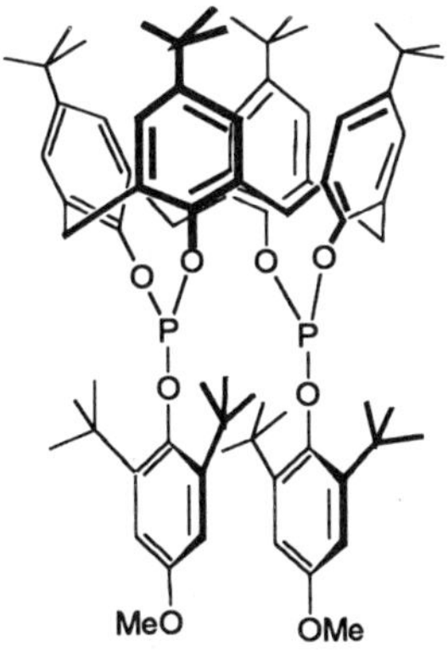 **38**

compared to its branched isomer 2-methyloctanal, which is the highest selectivity observed to date. The two sterically demanding 2,6-di-*tert*-butylphenoxy substituents explain why even small molecules such as 1-octene are hindered in their complexation to the metal centre. Accordingly, the less sterically demanding 1-octyl-rhodium intermediate is strongly preferred to its 2-octyl-rhodium regioisomer. Molecular modeling calculations showed that replacement of the two *tert*-butyl substituents of the aryloxy group in **38** with isopropyl and methyl groups results in a gradual opening of the pincers-shaped structure of the ligand. On this basis, complexes with isopropyl and methyl substituents in the aryloxy groups were predicted and actually found to be more active than **38**, while maintaining a relatively high regioselectivity.

Highly efficient calix[4]arene based di- and trinuclear Zn^{II} and Cu^{II} complexes with phosphodiesterase activity have been recently developed by the group of Reinhoudt.[39-43] The upper rim of calixarene **39**, blocked in the *cone* conformation by proper alkylation of the lower rim hydroxyls, was used by these authors as a flexible platform for the preorganization of up to four catalytic groups. Their di- and trinuclear complexes, together with mononuclear model compounds, are shown in **40-46**. HPNP and EPNP are the RNA and DNA model substrates, respectively.

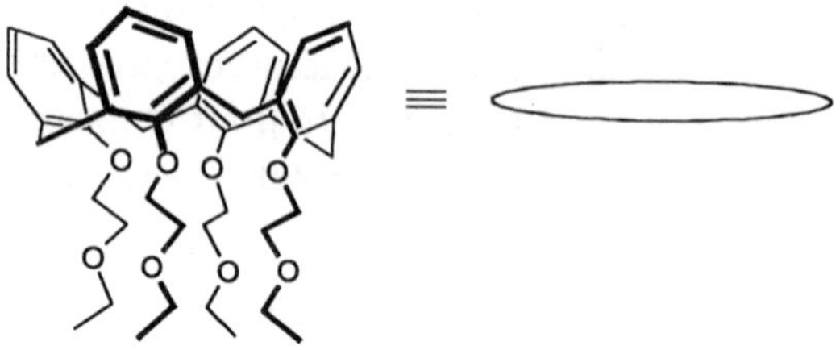

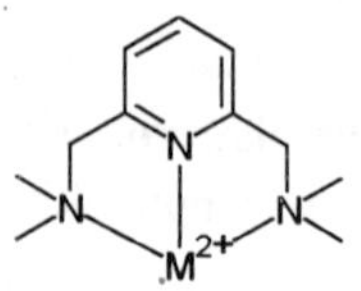

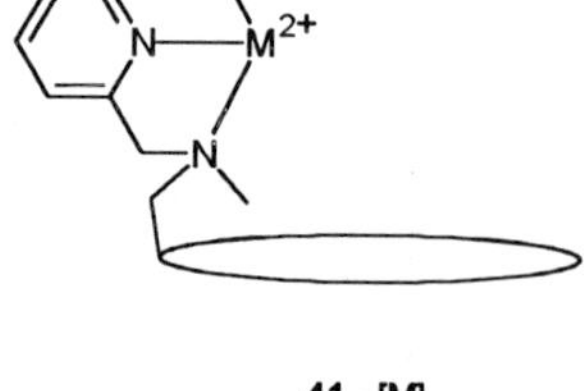

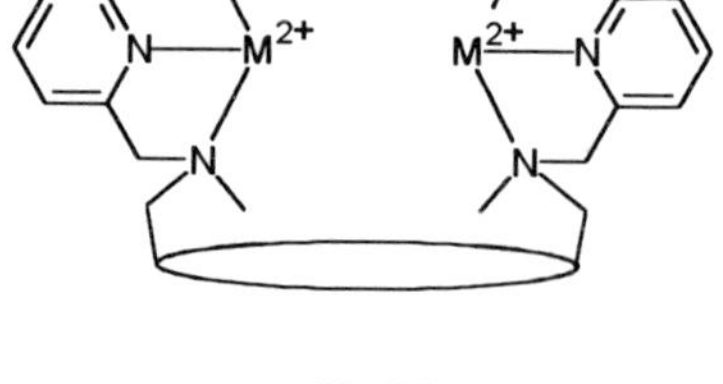

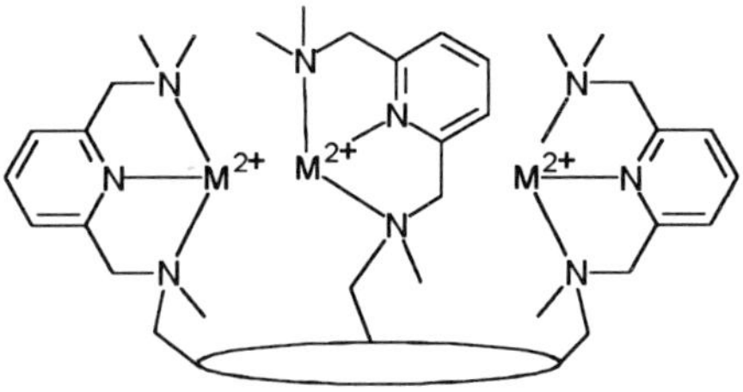

43 - [M]₃ 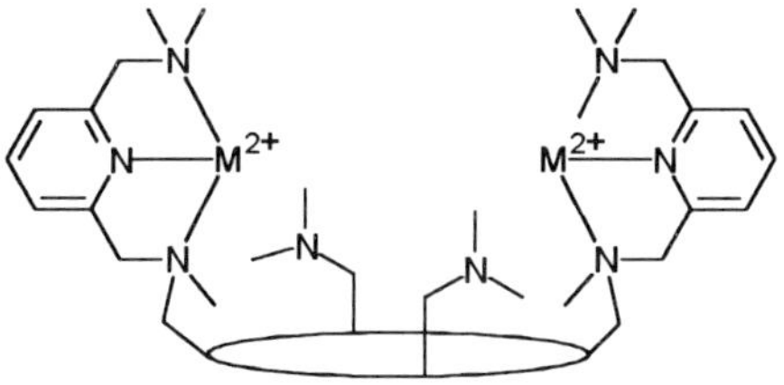   **44 - [M]₂**

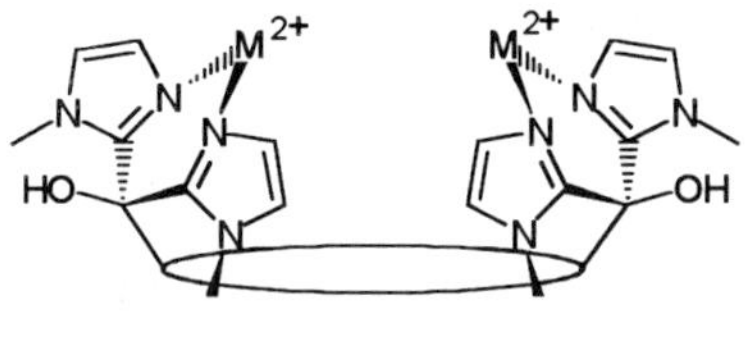

45 - [M]₂ 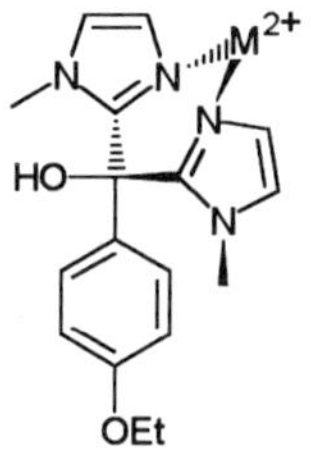   **46 - [M]** 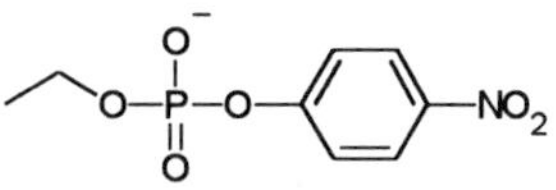   EPNP

The intramolecular transesterification of HPNP (Eq. 10.18) in MeCN-H₂O 1:1, 25 °C, pH 7, was accelerated by a factor of 23,000 in the presence of 0.48 mM of the dinuclear complex **42-[Zn]₂**.[39] A 4-fold excess of HPNP was completely cleaved without loss of activity in the time-course of reaction, showing that turnover catalysis occurs without significant product inhibition. In agreement with the two-step mechanism of Scheme 10.1, Michaelis-Menten kinetics were observed, from which the remarkably high value of 5.5×10^4 M^{-1} was obtained for the binding constant K. Since no saturation kinetics were observed with the mononuclear complex **41-[Zn]**, the unusually high binding constant of **42-[Zn]₂** was taken as evidence of a synergetic action of the two metals, most likely favoured by the adaptable calix[4]arene scaffold. The catalytic activity of **41-[Zn]** is 50 times lower than that of **42-[Zn]₂**, but 6 times higher than that of **40-[Zn]**, thus suggesting a

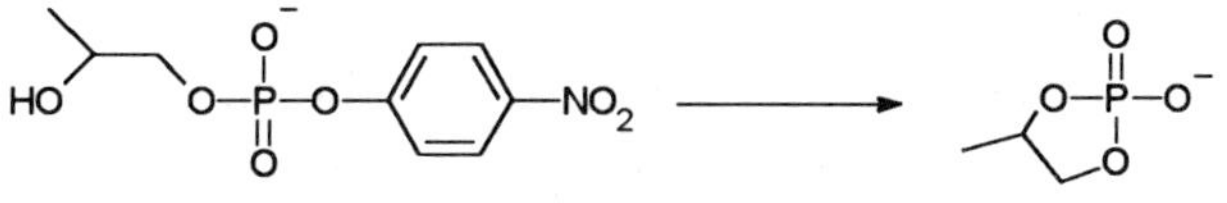

HPNP   (10.18)

47

possible contribution from the calixarene cavity to the catalysis. Complex **42-[Zn]$_2$** is not catalytically active in the hydrolysis of diethyl *p*-nitrophenyl phosphate, ethyl *p*-nitrophenyl phosphate (EPNP), and *p*-nitrophenyl phosphate, which shows the crucial role played by the β-hydroxyl of HPNP in the catalysis. The suggested mechanism for the **42-[Zn]$_2$** catalysed hydrolysis of HPNP is one in which one of the Zn^{II} ions coordinates the phosphate and the other is involved in activation of the nucleophilic hydroxyl, as depicted in **47**.

The two unsubstituted positions at the upper rim of **42-[Zn]$_2$** were used to introduce the two basic centres of **44-[Zn]$_2$**,[43] with the idea that a neighbouring general base could assist deprotonation of the nucleophilic hydroxyl. The kinetics strongly suggest that one of the amino group indeed acts as a general base in the catalytic process, but the steric hindrance of the additional pendant groups causes the activity of **44-[Zn]$_2$** to be lower than that of **42-[Zn]$_2$**.

For the purpose of designing a Cu^{II}-based dinuclear phosphodiesterase the calix[4]arene ligand **45**, containing chelating bisimidazolyl groups, is far superior to ligand **42**.[40] The dinuclear complex **45-[Cu]$_2$** catalyses with turnover and according to Michaelis-Menten kinetics not only the transesterification of the RNA model HPNP (rate enhancement 1.0×10^4, 35% EtOH, pH 6.2), but also, unlike the dinuclear complex **42-[Zn]$_2$**, the hydrolysis of the DNA model EPNP (rate enhancement 2.7×10^4, 35% EtOH, pH 6.4). Comparison of rate data obtained in the presence of the mononuclear catalyst **46-[Cu]** evidences the efficient synergetic action of the two metal ions in **45-[Cu]$_2$** both in the formation of the catalyst-substrate complex and in the subsequent transformation of the bound substrate. The suggested mechanism for the transesterification of HPNP (**48**) and

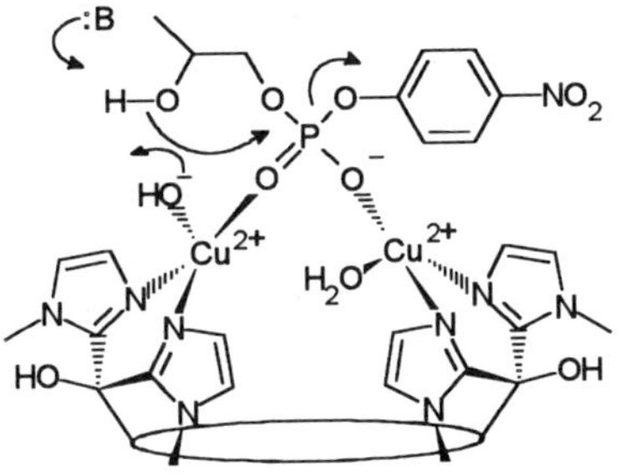

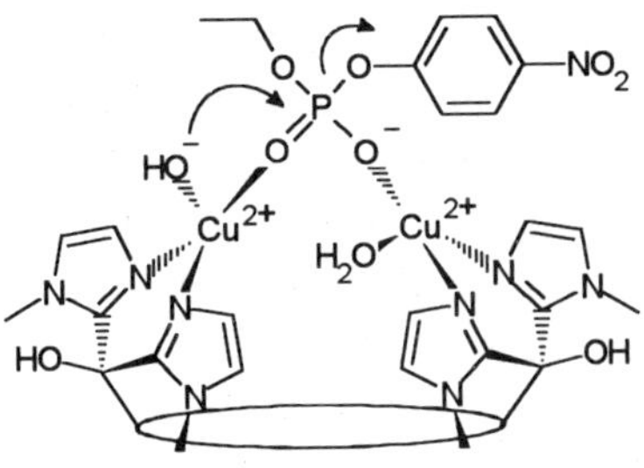

48 **49**

hydrolysis of EPNP (**49**) involves double Lewis acid activation in combination with bifunctional catalysis.

Catalytic studies with M-calixarenes have been recently extended to the trinuclear case.[41] Compared with the dinuclear complex **42**-[Zn]$_2$, the third metal centre of the trinuclear complex **43**-[Zn]$_3$ decreases the affinity toward HPNP by a 45-fold, but causes a 3-fold increase in catalytic rate. This means that **43**-[Zn]$_3$ is more active than **42**-[Zn]$_2$ under conditions close to saturation, but less active in more dilute solutions. In the trinuclear complex the three metal centres are clearly involved in the catalysis. It has been suggested that whereas two adjacent metal ions provide double Lewis acid activation to phosphate, the third metal ion facilitates deprotonation of the β-hydroxyl either by coordination to the oxygen atom or via a bound hydroxide ion. Even more importantly, **43**-[Zn]$_3$ catalyses the cleavage of the phosphate diester bond in RNA dinucleotides 3',5'-NpN.[42] As shown in Table 10.2, the trinuclear Cu^{II}-complex **43**-[Cu]$_3$ is much less effective, but a slightly enhanced activity is seen with the heterotrinuclear complex

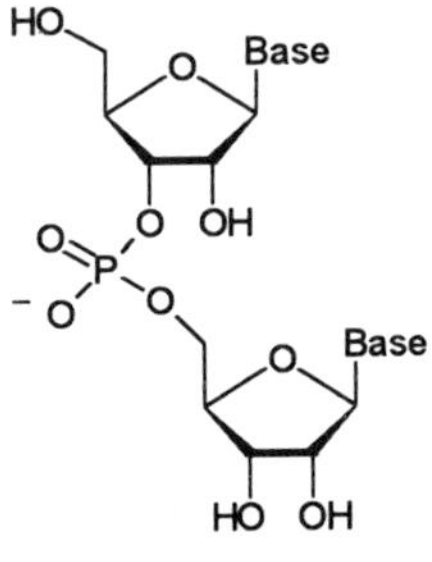

3',5' - NpN

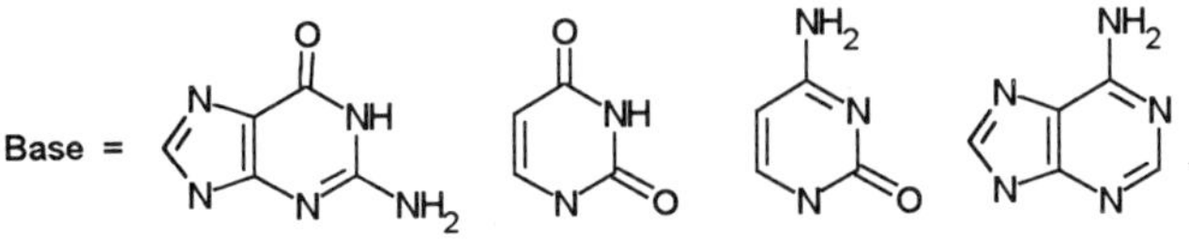

43-[Zn]₂[Cu]. Nucleobase specificity is the most interesting feature emerging from the data. The most reactive nucleotides GpG and UpU have an acidic amide NH that can be deprotonated, and the resulting nitrogen anion can bind strongly to Zn^{II}. Thus, one of the Zn^{II} centres in the trinuclear complex might act as a binding unit, whereas the two remaining metals may bind and activate the phosphate by double Lewis acid coordination (**50**). Trinuclear complexes **43-[Zn]₃** and **43-[Zn]₂[Cu]** are among the most active synthetic phosphodiesterases reported to date. Preliminary experiments with a synthetic 24-mer RNA oligonucleotide open prospects for sequence selective cleavage of RNA.

Table 10.2. First-order rate constants ($k_{obs}/10^5$ s^{-1}) for the cleavage of RNA dinucleotides 3',5'-NpN catalysed by trinuclear complexes **43-[M]₃** in 35% EtOH, pH 8.0 at 50 °C.[a]

3',5'-NpN	43-[Zn]₃	43-[Cu]₃	43-[Zn]₂[Cu]
GpG	72	28	88
UpU	8.5	1.2	13
CpC	6.1	1.9	7.1
GpA	4.6	-[b]	5.9
ApG	2.7	-[b]	2.4
ApA	0.44	0.47	0.46

a) [3',5'-NpN]=0.09 mM; [**43-M₃**]=0.9 mM.
b) not determined.

10.6. References

1. Funada, H.; Nakamoto, Y.; Ishida, S., *Polym. Prepr., Jpn.* **34**, (1985), 368.
2. a) Taniguchi, H. and Nomura, E., *Chem. Lett.*, (1988), 1773-1776; b) Taniguchi, H.; Otsuji Y.; Nomura E., *Bull. Chem. Soc. Jpn.* **68**, (1995), 3563-3567.
3. a) Nomura, E; Taniguchi, H.; Kawaguchi, K.; Otsuji, Y., *Chem. Lett.*, (1991), 2167-2170; b) Nomura E.; Taniguchi, H.; Kawaguchi, K.; Otsuji Y., *J. Org. Chem.* **58**, (1993), 4709-4715.
4. Nomura E.; Taniguchi, H.; Otsuji Y., *Bull. Chem. Soc. Jpn.* **67**, (1994), 792-799.
5. Nomura E.; Taniguchi, H.; Otsuji Y., *Bull. Chem. Soc. Jpn.* **67**, (1994), 309-311.
6. Okada, Y.; Sugitani, Y.; Kasai, Y.; Nishimura, J., *Bull. Chem. Soc. Jpn.* **67**, (1994), 586-588.
7. Araki, K.; Yanagi, A.; Shinkai, S., *Tetrahedron* **49**, (1993), 6763-6772.
8. Harris, S.J.; Kinahan, A.M.; Meegan, M.J.; Prendergast, R.C., *J. Chem. Research (S)*, (1994), 342-343.
9. Shimizu, S.; Kito, K.; Sasaki, Y.; Hirai, C., *J. Chem. Soc. Chem. Commun.*, (1997), 1629-1630.
10. Shinkai, S.; Mori, S.; Koreishi, H.; Tsubaki, T.; Manabe, O., *J. Am. Chem. Soc.* **108**, (1986), 2409-2416.
11. a) Moras, D. *et al.*, *J. Biol. Chem.* **250**, (1975), 9137-9162; b) Branlant, G. *et al.*, *Eur. J. Biochem.* **129**, (1982), 437-446.
12. Gutsche, C.D. and Alam, I., *Tetrahedron* **44**, (1988), 4689-4694.
13. Shinkai, S.; Shirahama, Y.; Tsubaki, T.; Manabe, O., *J. Chem. Soc., Perkin Trans. 1*, (1989), 1859-1860.
14. Shinkai, S.; Shirahama, Y.; Tsubaki, T.; Manabe, O., *J. Am. Chem. Soc.* **111**, (1989), 5477-5478.
15. Komiyama, M; Isaka, K.; Shinkai, S., *Chem. Lett.*, (1991), 937-940.
16. Pirrincioglu, N.; Zaman, F.; Williams, A., *J. Chem. Soc., Perkin Trans. 2*, (1996), 2561-2562.
17. Chawla, H.M. and Pathak, M., *Bull. Soc. Chim. Fr.* **128**, (1991), 232-243.
18. Schneider, H.-J. and Schneider, U., *J. Org. Chem.* **52**, (1987), 1613-1615.
19. Schneider, U. and Schneider, H.-J., *Chem. Ber.* **127**, (1994), 2455-2469.
20. Magrans, J.O.; Ortiz, A.R.; Molins, M.A.; Lebouille, P.H.P.; Sánchez-Quesada, J.; Prados P.; Pons, M.; Gago, F.; de Mendoza, J., *Angew. Chem. Int. Ed. Engl.* **35**, (1996), 1712-1715.
21. Cuevas, F.; Di Stefano, S.; Mandolini, L.; de Mendoza, J.; Prados, P., *to be published*.

22. Kuran, W.; Listos, T.; Abramczyk, M.; Dawidek, A. *J. Macromol. Sci.-Pure Appl. Chem.*, **A35**, (1998), 427-437.
23. Casolari, S.; Cozzi P.G.; Orioli, P.; Tagliavini, E.; Umani-Ronchi A., *J. Chem. Soc. Chem. Commun.*, (1997), 2123-2124.
24. Floriani, C. *Chem. Eur. J.* **5**, (1999), 19-23, and references cited therein.
25. Giannini, L.; Solari, E.; Zanotti-Gerosa A.; Floriani, C.; Chiesi-Villa, A.; Rizzoli, C., *Angew. Chem. Int. Ed. Engl.* **35**, (1996), 85-87.
26. Giannini, L.; Caselli, A.; Solari, E.; Floriani, C.; Chiesi-Villa, A.; Rizzoli, C.; Re, N.; Sgamellotti, A. *J. Am. Chem. Soc.* **119**, (1997), 9709-9719.
27. Castellano, B.; Solari, E.; Floriani, C.; Re, N.; Chiesi Villa, A.; Rizzoli, C. *Chem. Eur. J.* **5**, (1999), 722-737.
28. Giannini, L.; Guillemot, G.; Solari, E.; Floriani, C.; Re, N.; Chiesi-Villa, A.; Rizzoli, C. *J. Am. Chem. Soc.* **121**, (1999), 2797-2807.
29. Giannini, L.; Solari, E.; Dovesi, S.; Floriani, C.; Re, N.; Chiesi-Villa, A.; Rizzoli, C. *J. Am. Chem. Soc.* **121**, (1999), 2784-2796.
30. Cacciapaglia, R. and Mandolini, L., *Chem. Soc. Rev.* **22**, (1993), 221-231.
31. Cacciapaglia, R.; Casnati, A.; Mandolini L.; Ungaro, R., *J. Chem. Soc. Chem. Commun.* (1992), 1291-1293.
32. Cacciapaglia, R.; Casnati, A.; Mandolini L.; Ungaro, R., *J. Am. Chem. Soc.* **114**, (1992), 10956-10958.
33. Baldini, L.; Bracchini, C.; Cacciapaglia, R., Casnati, A.; Mandolini, L.; Ungaro, R., *to be published.*
34. Cacciapaglia, R.; Casnati, A.; Mandolini, L.; Ungaro, R., *unpublished results.*
35. Böhmer, V.; Cacciapaglia, R.; Mandolini, L.; Vogt, W., *unpublished results.*
36. Cacciapaglia, R.; Mandolini L.; Arnecke, R.; Böhmer, V; Vogt, W., *J. Chem. Soc. Perkin Trans. 2*, (1998), 419-423.
37. Loeber, C.; Wieser, C.; Matt, D.; De Cian, A.; Fischer, J.; Toupet, L., *Bull. Soc. Chim. Fr.* **132**, (1995), 166-177.
38. Paciello R.; Siggel, L.; Röper, M., *Angew. Chem. Int. Ed. Engl.* **38**, (1999), 1920-1923.
39. Molenveld, P.; Kapsabelis, S.; Engbersen, J.F.J; Reinhoudt, D.N., *J. Am. Chem. Soc.* **119**, (1997), 2948-2949.
40. Molenveld, P.; Engbersen, J.F.J; Kooijman, H.; Spek, A.L.; Reinhoudt, D.N., *J. Am. Chem. Soc.* **120**, (1998), 6726-6737.
41. Molenveld, P.; Stikvoort, W.M.G.; Kooijman H.; Spek, A.L.; Engbersen, J.F.J.; Reinhoudt, D.N., *J. Org. Chem.* **64**, (1999), 3896-3906.
42. Molenveld, P.; Engbersen, J.F.J.; Reinhoudt, D.N., *Angew. Chem. Int. Ed. Engl.*, in press.
43. Molenveld, P.; Engbersen, J.F.J.; Reinhoudt, D.N., *Eur. J. Org. Chem.* **64**, in press.

SUBJECT INDEX

A

Ab initio calculations 151, 156
Acceptor number 132
Acetate anion 117, 137
Acetylcholine 86-91, 96, 98-102,
    105-106, 249
Acid-catalysed condensation 3
Actinides 78
Activated amides 126
Adsorption, on thin films 184, 192
Air-water interface 174, 185, 197
Alkali metal ions 27, 63
Alkaline-earth metal ions 70, 72
Alkylamines 188
Allosteric effect 136
Ammonium ions 27, 186
Amphiphilic calix[n]arenes 173, 186
Anion
    binding sites 113
    coordination 111
    ionic radii 112
    pollutants 140
    receptors 111-114, 116-123
Antenna chromophore 78
Antenna effect 74
Aromatic guests 189, 193
Arrays, supramolecular 183
Artificial
    acetylcholinesterase 249
    esterases 247, 249
    phosphodiesterases 258, 260, 262
    ribonuclease 247
Au-S bonds 179-180, 189
Azobenzene chromophores 179

B

β-methyl glucopyranoside 227
Base-catalysed condensation 2
Basket-like calix[6]arene 95
Benzene/p-Xylene pair 236
Benzoate anion 121, 126, 137
Bifunctional catalysis 261
Bifunctional receptors 133, 135, 206
Biological recognition 85
Biologically important cations 80
Biphasic kinetics 255
Bis calixarenes 69, 127
Bis-roof structure 153
Bowl-shaped cavitands 101
Boxes 218-222
Bridged calixarenes 26, 65, 95-98, 104, 116,
    146
Bridged tetrahomo-dioxacalix[4]arene 102
Broken-chain mechanism 20
Burst kinetics 255

C

C_{60} 51, 53, 155-156, 188, 226
C_{70} 51, 53, 155-156
Cadmium sulfide 197
Cages 54, 155
Calix[4]arene podands 64
Calix[4]arene polyimides 194
Calix[4]arene tetraacids 73
Calix[4]arene tetramides 70
Calix[4]arene triamides 76
Calix[4]arene-biscrown (see
    Calixbiscrowns)

265

Calix[4]arene-crown ethers (see Calixcrowns)
Calix[4]tube 69
Calix[6]arene tricarboxylic acid 204
Calixarene bis-cyanurates 221
Calixarene bis-melamine 221
Calixarene ferrocene 131
Calixarene monoanions 150
Calixbiscrowns 69, 97, 148
Calixcrowns 26, 65, 95-98, 145, 146, 148, 188, 254, 256
Calixpyrroles 128
Calixquinones 71, 163
Calixresorcarenes 2, 87, 132, 175-189, 225, 228, 248
Calixspherands 24
Calorimetric measurements 93
Capsules 235
Carboxylate anions 117, 137
Carceplex 55, 198
Carceroisomerism 55
Catalysis 241-264
Catalytic receptors 245-250
Cation transport 66
Cation-π interaction 85, 87, 91, 92, 93, 97, 99, 147-155, 210
Cavitands 176, 189, 192, 226
CH-π interactions 91, 92, 144, 146, 160, 165, 189, 227
Charged anion receptors 116
Charged hydrogen bonds 233
Chiral calixarenes 47, 50, 188, 225, 251
Chiral capsules 214
Chiral guest 213, 236
Chloride binding 119
Choline 87, 88, 91
Chromatography 67

Chromogenic receptors 72
CMPO 79
Coalescence temperature, capsules 232
Cobaltocenium calix[4]arene 116
Columnar polymers 160
Combinatorial libraries 223
Competitive
 complexation 211
 crystallisation 144
 inhibition 246, 250
Complexation catalysis 247
Complexation Induced Shifts (CIS) 43, 87-89, 92, 98, 102, 103
Computational studies (see Molecular Modeling)
Conformation
 calix[4]arenes 5, 16, 62, 65, 67
 calix[4]diquinones 28, 64
 calix[5]arenes 17
 calix[6]arenes 18
 nomenclature 1-5
 on the surface 176
 symbolic representation 9
Colloidal silica 182, 185
Conformational inversion 5, 20, 23
Continous-chain mechanism 20
Cooperativity, hydrogen bonding 237
Coordinate driver method 21
Copper(I) ion 132
Covalent encapsulation 231
Crosslinked monolayers 191
Cryptands 114
Cryptophanes 93
Crystal engineering 158-167, 225
Cs^+/Na^+ selectivity 65, 66
Cyclophanes 85, 93
Cyclotriveratrylene (CTV) 123

D

Data storage materials 199
De Mendoza rule 7
Deep cavity cavitands 189, 234
Dialkoxycalix[4]arenes 46
Diastereomeric complexes 101
Dihydrogen phosphate 134
Dimer size 234
Dimeric capsule 226
1,3-Dimethoxycalix[4]arenes 63
1,3-Dimethoxycalixcrowns 63
Dinuclear complexes 150
Ditopic receptors 133
Divalent cations, calculations 28
DNA cleavage 259
Double calixarenes 94, 97, 103, 105, 215
Double container 231
Double displacement mechanism 255
Doubly bridged calix[4]arene 104
Dynamic
    combinatorial library 221
    nanostructures 223
    stereochemistry 232

E

Electrochemical data 72, 118
Electrochemical recognition 121
Electrochemiluminescence properties 197
Electrostatic interactions 87, 93, 245, 247
Endo-calix catalytic process 243, 245
Energy barrier 232
Enzyme catalysis 241
ESI-MS 210, 217
Ester cleavage 246-249
Europium complexes 75

Exo-calix catalytic process 247
Extraction 73

F

Fluorescence
    quenching 89
    monolayers 197
    regeneration 89
    studies 121
Fluoride anion 128
Free energy calculations 27
Free methyl quantum rotors 146
Fullerenes 51, 155-158

G

Gas phase complexation 191
General acid catalysis 256
General base catalysis 261
Glucoronolactone 187
Gold surfaces 189
Gold-thiol monolayers 180
Graphite surface 223
Grazing angle FTIR 180

H

Halide anions 27, 117, 119, 123-125, 139
Halogenated hydrocarbons 189
Hard Soft Acid Base (HSAB) 72
Heterocalixarene analogues 128
Hetero-dimers 205, 209, 211, 232, 234
Heteroditopic receptors 94, 105
Hexameric assembly 228
Historical Notes
    anion binding 114
    calixarenes 1

Hole-size relationship 94
Homo-capsules 214
Homooxacalixarenes 51, 58, 93, 94, 101
HPLC separation 130
Hydration,NADH 245
Hydroformylation, olefins 257
Hydrogen bond anion receptors 124
Hydrogen bonded dimers 204, 207
Hydrogen bonding 113, 205-206, 218
Hydrolysis, phosphate esters 247
Hydrophobic effect 87, 93, 168, 174
Hyperpolarisabilities 195

I

Inelastic neutron scattering (INS) 146
Interdigitation 181
Iodine doping 197
Ion channel 185
Ion pairs 98, 133, 138
Ion Selective Electrodes (ISEs) 66, 68
Ion Sensitive Field Effect Transistors (ISFETs) 66
Ionic radii 112
Ionisable ligands 80

K

K^+/Na^+ selectivity 65
Kinetic stability, capsules 207, 211, 212
Koilands, Koilates 158

L

Langmuir isotherm 177
Langmuir-Blodgett (LB) films 173, 188, 193
Langmuir-Blodgett multilayers 177, 184
Lanthanide ion complexes 28, 74, 76, 77, 185
Layered structures 163-165
Lewis acid activation 249, 255, 261
Lewis acid centres 113, 115, 130
Linear "polycaps" 215
Lipid bilayer 185
Liquid-liquid extraction 67
Luminescent complexes 74, 77

M

MALDI-TOF mass spectrometry 220
Mass spectrometry 40, 209
McPC603 antibody 96, 105, 249
MD Simulations 13
Melamine-barbiturate 218
Metal cations 62-85, 175
Metalated calixarenes 122-123
Michaelis-Menten kinetics 241, 245, 248, 259, 260
Migratory insertion 252, 253
Molecular Dynamics 13
Molecular Mechanics (MM) calculations 13, 145
Molecular modeling 11, 13, 27-29, 75, 119, 156
Molecular networks 158
Molecular switching device 198
Monolayers 179, 184
Monolayers on silica 182
Monovalent cations 27, 28
Monte Carlo method 13
Multichelate effect 86
Multilayers 157, 176, 196

N

Na$^+$/K$^+$ selectivity 68
Na$^+$ complexation 206
Nano-particles 198
Nanostructures 222
Neutral anion receptors 124-140
Neutral guests 38-57, 186, 193, 230
Neutron reflectometry 180
Nitrate anion 123
p-Nitrocalix[4]arenes 194
N-methyl-4-picolinium 105, 205
N-methylpyridinium cation 94, 95, 100, 103, 104, 105, 107, 205
N-methylquinuclidinium ion 88, 98, 102, 217
N,N,N-trimethylanilinium cation 90, 92, 93, 98, 102, 246
NMR, capsules 212
NMR patterns, calix[4]arenes 6
NMR titrations 138
NOESY 213
Nomenclature 1
Nonlinear Optical Properties (NLO) 194, 196
Nuclear industry 123
Nuclear waste 66, 78
Nucleotides 130

O

Octahedral assembly 228
Octahedral-cubic assembly 227
Oligonucleotides 130
Optical sensors 194
Ordered monolayers 179

Organic
  amines 188
  cations 86
  molecules
  pollutants 35-36, 188-189, 230, 192
  vapours 49, 188-189, 192
Organised bilayers 180
Organometallic bridges 159

P

π-Metalated calixarenes 122-123
π-π- Interactions 52, 192
Perchloroethylene 188
Permeation-selective materials 174
Phase transfer catalysis 242-245
Phosphate binding 119
Phosphatidylcholine 105
Phosphine complexes 132
Phosphine oxide groups 78
Phosphocholine binding site 96, 249
Phosphodiesterase 261
Phosphonium cations 97
Ping-pong mechanism 255
Polyammonium macrocycles 115
Polycaps 216
Polymerization, monolayers 181
Polysaccharides 187
Porous layers 182, 193
Porphyrins 106, 206, 227
Potassium channels 70
Potassium ion 134
Potential energy calculations 145
Pressure-area isotherm 184
Propionate anion 137
Protonated amines 114
Pseudorotation 24

Pyroelectric effect 196
Pyrogallol tetramer 228

Q

Quartz Crystal Microbalances (QCM) 188-193
Quartz plates 183
Quaternary ammonium cations (see Quats)
Quats 27, 85-110

R

Radioactive waste treatment 80
Radioimmunotheraphy 80
Radionuclides 68
Reactive monolayer surface 190
Redox couple 116, 118
Regioselective alkylation 62
Resorcarenes (see calixresorcarenes)
Reversible bonds 203
Rhenium(I) calixarenes 130
Rigid cone calix[4]arenes 44, 97
Ring inversion 5, 47
RNA cleavage 262
RNA dinucleotide 262
Rod-like assemblies 223
Rosettes 218, 221
Ruthenium(II) calixarenes 120, 130

S

Sauerbray equation 48
Saturation kinetics 246, 247, 259
Second order supermolecules 236
Second-order nonlinearities 195
Second-sphere co-ordination 167

Selective bridging 95
Self-assembled monolayers 179, 190
Self-assembly 160, 203, 223
Self-inclusion 165
Semiconductor materials 197
Sensor chips 187
Sesquicalix[4]arene 105
Shapes of anions 112
Silver ion complexation 132, 154
SIMS 95
Solvent effect, anion binding 131
Solvent effects, quats binding 99
Spherical metal ions 62-64
SPR spectroscopy 190
Sr^{2+}/Na^+ selectivity 74
Steric effects, binding 189
Stilbenes 235
Structural properties, calixarenes 144-171
Sugar binding 76, 186, 187, 227
Sugar cluster 186
Sulfonatocalix[n]arenes 90, 92, 173, 245
Sulphonamide receptors 126
Supported Liquid Membranes (SLM) 66-68
Supramolecular
  arrays 183
  diastereomers 219
  helicity 221
  materials 172-202
Suzuki coupling 234
Synthesis of calixarenes 1
Synthetic ion channels 185

T

Terbium complexes 75, 78
Tetraaminocalix[4]arenes 207
Tetrahydroxy cavitands 229

Tetralkylammonium cations 88, 95, 97, 98, 103, 105
Tetramethoxycalix[4]arene 155
Tetramethylammonium ion 95, 162
Tetraurea self-assembly 207-215
Theoretical investigations 24, 144-171, 195
Thermodinamics, capsule formation 210
Thin films 172, 188-189
Thiol-resorcinarenes 187
Thiourea receptors 136
Three dimensional arrays 183
TM-SFM 223
Transition metals, π-complexes 122, 152-154
Transition state analogue 249
Transport studies 66, 79
Tricarbonylchromium complexes 152
Trifluoromethanol receptors 128
Trivalent cations 28
Tunnel effect 146
Turnover catalysis 250-260
Twistomers 232

U

Uranyl cation 28
Urea-hinges 211
Urea receptors 125, 136, 207

V

Vancomycin mimics 51
Vapour binding selectivity 188
Vitamin C 187

W

Water activation 249
Water soluble calixarenes 86, 90, 164, 245
Water surface 178
Water-air interface 174, 185, 197
Wolframium complex 161

X

XPS measurements 184, 198
X-ray analysis 209
X-ray reflectometry 177
Xylene isomers 189

Z

Zirconium-BINOL catalyst 251
Zn-porphyrin 206